Birkhäuser

Modern Birkhäuser Classics

Many of the original research and survey monographs in pure and applied mathematics published by Birkhäuser in recent decades have been groundbreaking and have come to be regarded as foundational to the subject. Through the MBC Series, a select number of these modern classics, entirely uncorrected, are being re-released in paperback (and as eBooks) to ensure that these treasures remain accessible to new generations of students, scholars, and researchers.

M. Ram Murty
V. Kumar Murty

Non-vanishing of *L*-Functions and Applications

Reprint of the 1997 Edition

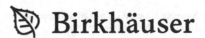 Birkhäuser

M. Ram Murty
Department of Mathematics and Statistics
Jeffery Hall, Queen's University
Kingston, ON, K7L 3N6
Canada

V. Kumar Murty
Department of Mathematics
University of Toronto
40, St. George Street
Toronto, ON M5S 2E4
Canada

ISBN 978-3-0348-0273-4   e-ISBN 978-3-0348-0274-1
DOI 10.1007/978-3-0348-0274-1
Springer Basel Dordrecht Heidelberg London New York

Library of Congress Control Number: 2011941445

Mathematics Subject Classification (2010): 11Mxx, 11M41, 11G40, 11R52, 11R42

© Springer Basel AG 1997
Reprint of the 1st edition 1997 by Birkhäuser Verlag, Switzerland
Originally published as volume 157 in the Progress in Mathematics series

This work is subject to copyright. All rights are reserved, whether the whole or part of the material is concerned, specifically the rights of translation, reprinting, re-use of illustrations, recitation, broadcasting, reproduction on microfilms or in other ways, and storage in data banks. For any kind of use, permission of the copyright owner must be obtained.

Printed on acid-free paper

Springer Basel AG is part of Springer Science+Business Media
(www.birkhauser-science.com)

Fernando Sunyer i Balaguer 1912–1967

* * *

This book has been awarded the Ferran Sunyer i Balaguer 1996 prize.

Each year, in honor of the memory of Ferran Sunyer i Balaguer, the Institut d'Estudis Catalans awards an international research prize for a mathematical monograph of expository nature. The prize-winning monographs are published in this series. Details about the prize can be found at
`http://www.iec.es/fsbprang.htm`

Previous winners include

- *Alexander Lubotzky*
  Discrete Groups, Expanding
  Graphs and Invariant Measures
  (vol. 125)
- *Klaus Schmidt*
  Dynamical Systems of Algebraic
  Origin (vol. 128)

Fernando Sunyer i Balaguer 1912–1967

Born in Figueras (Gerona) with an almost fully incapacitating physical disability, Fernando Sunyer i Balaguer was confined for all his life to a wheelchair he could not move himself, and was thus constantly dependent on the care of others. His father died when Don Fernando was two years old, leaving his mother, Doña Angela Balaguer, alone with the heavy burden of nursing her son. They subsequently moved in with Fernando's maternal grandmother and his cousins Maria, Angeles, and Fernando. Later, this exemplary family, which provided the environment of overflowing kindness in which our famous mathematician grew up, moved to Barcelona.

As the physician thought it advisable to keep the sickly boy away from all sorts of possible strain, such as education and teachers, Fernando was left with the option to learn either by himself or through his mother's lessons which, thanks to her love and understanding, were considered harmless to his health. Without a doubt, this education was strongly influenced by his living together with cousins who were to him much more than cousins for all his life. After a period of intense reading, arousing a first interest in astronomy and physics, his passion for mathematics emerged and dominated his further life.

In 1938, he communicated his first results to Prof. J. Hadamard of the Academy of Sciences in Paris, who published one of his papers in the Academy's "Comptes Rendus" and encouraged him to proceed in his selected course of investigation. From this moment, Fernando Sunyer i Balaguer maintained a constant interchange with the French analytical school, in particular with Mandelbrojt and his students. In the following years, his results were published regularly. The limited space here does not, unfortunately, allow for a critical analysis of his scientific achievements. In the mathematical community his work, for which he attained international recognition, is well known.

Don Fernando's physical handicap did not allow him to write down any of his papers by himself. He dictated them to his mother until her death in 1955, and when, after a period of grief and desperation, he resumed research with new vigor, his cousins took care of the writing. His working power, paired with exceptional talents, produced a number of results which were eventually recognized for their high scientific value and for which he was awarded various prizes. These honours not withstanding, it was difficult for him to reach the social and professional position corresponding to his scientific achievements. At times, his economic situation was not the most comfortable either. It wasn't until the 9th of December 1967, 18 days prior his death, that his confirmation as a scientific member was made public by the División de Ciencias, Médicas y de Naturaleza of the Council. Furthermore, he was elected only as "de entrada", in contrast to class membership.

Due to his physical constraints, the academic degrees for his official studies were granted rather belatedly. By the time he was given the Bachelor degree, he had already been honoured by several universities! In 1960 he finished his Master's degree and was awarded the doctorate after the requisite period of two years as a student. Although he had been a part-time employee of the Mathematical Seminar since 1948, he was not allowed to become a full member of the scientific staff until 1962. This despite his actually heading the department rather than just being a staff member.

His own papers regularly appeared in the journals of the Barcelona Seminar, *Collectanea Mathematica*, to which he was also an eminent reviewer and advisor. On several occasions, he was consulted by the Proceedings of the American Society of Mathematics as an advisor. He always participated in and supported guest lectures in Barcelona, many of them having been prepared or promoted by him. On the occasion of a conference in 1966, H. Mascart of Toulouse publicly pronounced his feeling of beeing honoured by the presence of M. Sunyer Balaguer, "the first, by far, of Spanish mathematicians".

At all times, Sunyer Balaguer felt a strong attachment to the scientific activities of his country and modestly accepted the limitations resulting from his attitude, resisting several calls from abroad, in particular from France and some institutions in the USA. In 1963 he was contracted by the US Navy, and in the following years he earned much respect for the results of his investigations. "His value to the prestige of the Spanish scientific community was outstanding and his work in mathematics of a steady excellence that makes his loss difficult to accept" (letter of condolence from T.B. Owen, Rear Admiral of the US Navy).

Twice, Sunyer Balaguer was approached by young foreign students who wanted to write their thesis under his supervision, but he had to decline because he was unable to raise the necessary scholarship money. Many times he reviewed doctoral theses for Indian universities, on one occasion as the president of a distinguished international board. The circumstances under which Sunyer attained his scientific achievements, also testify to his remarkable human qualities. Indeed, his manner was friendly and his way of conversation reflected his gift for friendship as well as enjoyment of life and work which went far beyond a mere acceptance of the situation into which he had been born. His opinions were as firm as they were cautious, and at the same time he had a deep respect for the opinion and work of others. Though modest by nature, he achieved due credit for his work, but his petitions were free of any trace of exaggeration or undue self-importance. The most surprising of his qualities was, above all, his absolute lack of preoccupation with his physical condition, which can largely be ascribed to the sensible education given by his mother and can be seen as an indication of the integration of the disabled into our society.

On December 27, 1967, still fully active, Ferran Sunyer Balaguer unexpectedly passed away. The memory of his remarkable personality is a constant source of stimulation for our own efforts.

Translated from Juan Augé: Fernando Sunyer Balaguer. *Gazeta Matematica*, 1.a Serie – Tomo XX – Nums. 3 y 4, 1968, where a complete bibliography can be found.

Satyam Jnanam Anantam Brahma

Table of Contents

Preface

This monograph brings together a collection of results on the non-vanishing of L-functions. The presentation, though based largely on the original papers, is suitable for independent study. A number of exercises have also been provided to aid in this endeavour. The exercises are of varying difficulty and those which require more effort have been marked with an asterisk.

The authors would like to thank the Institut d'Estudis Catalans for their encouragement of this work through the Ferran Sunyer i Balaguer Prize. We would also like to thank the Institute for Advanced Study, Princeton for the excellent conditions which made this work possible, as well as NSERC, NSF and FCAR for funding.

Princeton
August, 1996

M. Ram Murty
V. Kumar Murty

Introduction

Since the time of Dirichlet and Riemann, the analytic properties of L-functions have been used to establish theorems of a purely arithmetic nature. The distribution of prime numbers in arithmetic progressions is intimately connected with non-vanishing properties of various L-functions. With the subsequent advent of the Tauberian theory as developed by Wiener and Ikehara, these arithmetical theorems have been shown to be equivalent to the non-vanishing of these L-functions on the line $\mathrm{Re}(s) = 1$.

In the 1950's, a new theme was introduced by Birch and Swinnerton-Dyer. Given an elliptic curve E over a number field K of finite degree over $\mathbb{Q}$, they associated an L-function to E and conjectured that this L-function extends to an entire function and has a zero at $s = 1$ of order equal to the $\mathbb{Z}$-rank of the group of K-rational points of E. In particular, the L-function vanishes at $s = 1$ if and only if E has infinitely many K-rational points.

The analytic continuation of the L-series associated to E has now been established in the work of Wiles and his school for all elliptic curves which have semistable reduction at 3 and 5. So it now makes sense to talk about the values of the L-function of such curves for any point of the complex plane. In recent work of V. Kolyvagin, it was necessary to have a quadratic twist of a given elliptic curve whose L-function has a simple zero at $s = 1$.

This monograph is concerned with the non-vanishing of a general L-function, with special emphasis on classical Dirichlet L-functions, Artin L-functions and L-functions attached to modular forms.

The first technique to prove a theorem on non-vanishing arose in the work of Hadamard and de la Vallée Poussin. It is based on the simple trigonometric inequality

$$3 + 4\cos\theta + \cos 2\theta \geq 0.$$

It is remarkable that such a technique is capable of vast generalization. Theorem 1.2 of Chapter 1 is one such generalization. Theorem 4.1 of Chapter 3 is another. Both of these theorems have immediate applications to general questions concerning equidistribution.

The prime number theorem is a special case of the more general Chebotarev density theorem. In Chapters 1 and 2, we trace the development of these ideas and discuss in some detail analytic properties of Artin L-functions. The effective Chebotarev density theorem, which plays such an essential role in many questions of an arithmetic and diophantine nature, is also described.

In Chapter 3, we discuss a general formalism due to Serre. It becomes clear that questions of uniform distribution reduce to questions about analytic continuation of L-functions through the Tauberian theorems and an appropriate use of trigonometric inequalities.

The subject of modular forms and their associated L-functions has its origin in the works of Ramanujan and Hecke. After reviewing quickly some basic notions, we discuss in Chapter 4 the theme of non-vanishing of modular L-functions and their symmetric power analogues. This is done in the context of the Sato-Tate conjecture. We also discuss the application of these ideas to questions of oscillation of Fourier coefficients of cusp forms.

It is a 'folklore' conjecture that the classical Dirichlet L-function $L(s, \chi)$, associated to a Dirichlet character $\chi(\mathrm{mod}\, q)$ does not vanish at the central critical point $s = 1/2$. As of 1995, this is still unproved. In Chapter 5, we discuss this question from a variety of methods. First, one can consider averages such as

$$\sum_{\chi \bmod q} L(\frac{1}{2}, \chi) \quad \text{and} \quad \sum_{\chi \bmod q} |L(\frac{1}{2}, \chi)|^2.$$

By developing asymptotic formulas for these averages, one can obtain the existence of many characters χ for which $L(\frac{1}{2}, \chi) \neq 0$. To get stronger results, one considers not the above averages, but sums which are weighted with an auxilliary function. In work of Balasubramanian and K. Murty, it is shown that for each sufficiently large prime q, the number of Dirichlet characters $\chi(\mathrm{mod}\, q)$ such that $L(1/2, \chi) \neq 0$ is at least $\geq (.04)\phi(q)$. This is the content of Theorem 5.1 of Chapter 5. The methods are involved and based on the study of the averages

$$\sum_{\chi \bmod q} L(\frac{1}{2}, \chi) M_z(\frac{1}{2}, \chi)$$

where $M_z(s, \chi)$ is a Dirichlet polynomial which 'mollifies' the L function.

The method of averages to prove non-vanishing of L-functions is developed in Chapter 5 in the context of Dirichlet L-functions, and in Chapter 6 in the context of L-functions of modular forms. The main result of Chapter 6 shows that for a holomorphic modular form f which is a newform of weight 2, there is a quadratic character χ such that the twisted L-function $L(s, f, \chi)$ does not vanish at the central critical point.

The method of averages can be summarized by considering the following general problem. Suppose we are given a Dirichlet series

$$f(s) = \sum_{n=1}^{\infty} \frac{a_n}{n^s}$$

which converges absolutely in some half plane and extends to an analytic function in the region $\mathrm{Re}(s) > c$. Suppose further that all the twists

$$f(s,\chi) = \sum_{n=1}^{\infty} \frac{a_n \chi(n)}{n^s}$$

by Dirichlet characters $\chi(\mathrm{mod}\, q)$ have the same property. Given $s_0 \in \mathbb{C}$ such that $\mathrm{Re}(s_0) > c$, does there exist $\chi(\mathrm{mod}\, q)$ such that $f(s_0, \chi) \neq 0$? To answer this, it is natural to study

$$\sum_{\chi \bmod q} f(s_0, \chi)$$

and determine its asymptotic behaviour. More generally, one can study

$$\sum_{q \leq Q} \sum_{\chi \bmod q} c_\chi f(s_0, \chi).$$

Such a study was necessary in the recent work of V. Kolyvagin on the Birch and Swinnerton-Dyer conjecture. In his situation $f(s)$ was the L-function of a modular elliptic curve, $s_0 = 1$ and $c_\chi = 0$ unless χ is of order 2. We derive asymptotic formulas for such sums in Chapter 6.

This technique has been amplified and expanded in many works such as that of Iwaniec, Luo-Rudnick-Sarnak, Barthel-Ramakrishnan, K. Murty-Stefanicki, and Y. Zhang.

There are at least two more important techniques of non-vanishing of L-functions that are not discussed in this book. One is the method of Rohrlich which can be termed 'Galois theoretic'. The other is the 'automorphic method' of Bump, Friedberg and Hoffstein.

The important topic of general automorphic L-functions is not touched in this monograph. In this connection, we refer the reader to the monograph of Gelbart and Shahidi.

Finally, in Chapter 7, we discuss Selberg's conjectures concerning Dirichlet series with Euler products and functional equations. These conjectures imply that no element of the Selberg class vanishes on the line $\mathrm{Re}(s) = 1$. Most likely, the Selberg class coincides with the class of automorphic L-functions. An intriguing pathway of research is to compare and contrast these two points of view.

Chapter 1
The Prime Number Theorem
and Generalizations

§1 The Prime Number Theorem

It was a century ago that Jacques Hadamard and Charles de la Vallée Poussin proved (independently) the celebrated prime number theorem. If $\pi(x)$ denotes the number of primes up to x, the theorem states that

$$\lim_{x \to \infty} \frac{\pi(x)}{x/\log x} = 1.$$

Their method had its origins in a fundamental paper of Riemann written by him in 1860. That paper outlines a 'program' for proving the prime number theorem. It begins by introducing the ζ function which is defined for $\mathrm{Re}(s) > 1$ as

$$\zeta(s) = \sum_{n=1}^{\infty} \frac{1}{n^s}.$$

Riemann then proceeds to show that $(s-1)\zeta(s)$ extends to an entire function and satisfies a functional equation

$$\pi^{-s/2}\Gamma(\frac{s}{2})\zeta(s) = \pi^{-(1-s)/2}\Gamma(\frac{1-s}{2})\zeta(1-s).$$

In addition, $\zeta(s)$ can be written as an infinite product over the prime numbers p:

$$\zeta(s) = \prod_p \left(1 - \frac{1}{p^s}\right)^{-1} \qquad \mathrm{Re}(s) > 1. \tag{1}$$

This equality is an analytic reformulation of the fact that every natural number is a product of prime numbers in an (essentially) unique way. Because the product is absolutely convergent in $\mathrm{Re}(s) > 1$, equation (1) also reveals that $\zeta(s)$ does not vanish in this half-plane. Earlier, Euler[1] had noticed that unique factorization of the integers could be written in this way as well as the functional equation for the zeta function, but he treated $\zeta(s)$ as a function of a real variable. Riemann emphasized that many intricate questions about the distribution of prime numbers can, by virtue of the above identity, be translated into complex analytic questions involving the ζ function.

It took several decades to vindicate Riemann's approach and put it on rigorous footing. Many new ideas of complex analysis were discovered and developed as a consequence. By the time Hadamard and de la Vallée Poussin completed their proof, there was a general method in place for tackling all such questions. At the heart of their proof is the fact that the zeta function does not vanish on the line $\mathrm{Re}(s) = 1$. As it later transpired, this non-vanishing theorem is equivalent to the prime number theorem.

It is interesting to note that Hadamard, in his characteristic humility, writes, "Stieltjes avait démontré que tous les zéros imaginaires de $\zeta(s)$ sont (conformément aux prévisions de Riemann) de la forme $1/2 + it$, t étant réel; mais sa démonstration n'a jamais été publiée. Je me propose simplement de fair voir que $\zeta(s)$ ne saurait avoir de zero dont la partie réele soit égale à 1." (Oeuvres, p. 183).

We now understand this in a better light. The dominant theme that arises from the papers of Riemann, Hadamard, and de la Vallée Poussin is the following. Suppose we are given a sequence of complex numbers a_n and we would like to know the behaviour of

$$\sum_{n \leq x} a_n.$$

The idea is to study the associated Dirichlet series

$$f(s) = \sum_{n=1}^{\infty} \frac{a_n}{n^s}$$

as a function of a complex variable and infer from the analytic properties the desired behaviour of the summatory function. Indeed, suppose that all the a_n's are bounded by some constant C. Then the associated Dirichlet series defines an analytic function for $\mathrm{Re}(s) > 1$. Suppose further that the series can be continued analytically to $\mathrm{Re}(s) > 1 - \delta$ where $\delta > 0$. Beginning with the fundamental line integral

$$\frac{1}{2\pi i} \int_{c-i\infty}^{c+i\infty} \frac{x^s}{s} ds = \begin{cases} 1 & \text{if } x > 1 \\ 1/2 & \text{if } x = 1 \\ 0 & \text{if } x < 1 \end{cases}$$

---

[1]   Equation (1) is referred to as an Euler product. More general Euler products will be introduced later in the chapter.

for any $c > 0$, we easily see by term by term integration,

$$\sum_{n \leq x} a_n = \frac{1}{2\pi i} \int_{c-i\infty}^{c+i\infty} f(s) \frac{x^s}{s} ds$$

when x is not an integer. Here c is chosen so that $f(s)$ converges absolutely on $\mathrm{Re}(s) = c$. We can now invoke methods of contour integration and attempt to infer something about the behaviour of the sum in question. For instance, in the case under discussion, if we assume in addition that

$$f(s) = \mathbf{O}(\log^A(|s| + 2)),$$

for some constant $A > 0$, then it is it is not difficult to deduce that

$$\sum_{n \leq x} a_n = \mathbf{O}(x^\theta)$$

for any $\theta > 1 - \delta$.

Over the subsequent decades, the techniques and methods have been streamlined and made elegant and efficient, notably through the work of Hardy, Littlewood, Ikehara and Wiener. The following theorem represents the quintessence of their work and goes under the parlance of the (Wiener-Ikehara) Tauberian theorem.

Theorem 1.1. *Let $f(s) = \sum_{n=1}^{\infty} a_n/n^s$ be a Dirichlet series. Suppose there exists a Dirichlet series $F(s) = \sum_{n=1}^{\infty} b_n/n^s$ with positive real coefficients such that*
(a) *$|a_n| \leq b_n$ for all n;*
(b) *the series $F(s)$ converges for $\mathrm{Re}(s) > 1$;*
(c) *the function $F(s)$ (respectively $f(s)$) can be extended to a meromorphic function in the region $\mathrm{Re}(s) \geq 1$ having no poles except (respectively except possibly) for a simple pole at $s = 1$ with residue $R \geq 0$ (respectively r).*
Then

$$A(x) := \sum_{n \leq x} a_n = rx + \mathbf{o}(x)$$

as $x \to \infty$. In particular, if $f(s)$ is holomorphic at $s = 1$, then $r = 0$ and $A(x) = \mathbf{o}(x)$ as $x \to \infty$.

Remark. Note that we can equally deduce that

$$\sum_{n \leq x} b_n = Rx + \mathbf{o}(x)$$

as $x \to \infty$.

We relegate the proof of this theorem to later in this chapter. For the moment, we will quickly proceed to deduce the prime number theorem from the non-vanishing of $\zeta(s)$ on $\mathrm{Re}(s) = 1$.

Let us begin by observing that for any Dirichlet series,

$$\sum_{n=1}^{\infty} \frac{a_n}{n^s} = \sum_{n=1}^{\infty} \frac{A(n) - A(n-1)}{n^s}$$

$$= \sum_{n=1}^{\infty} A(n) \left(\frac{1}{n^s} - \frac{1}{(n+1)^s} \right)$$

$$= s \sum_{n=1}^{\infty} A(n) \int_{n}^{n+1} \frac{dx}{x^{s+1}}$$

$$= s \int_{1}^{\infty} \frac{A(x)}{x^{s+1}} dx.$$

In particular,

$$\zeta(s) = s \int_{1}^{\infty} \frac{[x]}{x^{s+1}} dx = \frac{s}{s-1} - s \int_{1}^{\infty} \frac{\{x\}}{x^{s+1}} dx \qquad (2)$$

where $[x]$ denote the greatest integer less than or equal to x and $\{x\} = x - [x]$. Since the fractional part of x is less than 1, the integral

$$\int_{1}^{\infty} \frac{\{x\}}{x^{s+1}} dx$$

converges for $\operatorname{Re}(s) > 0$. Hence, we obtain an analytic continuation of $(s-1)\zeta(s)$ in this half-plane. Furthermore, $\zeta(s)$ has a simple pole at $s = 1$ with residue 1.

By taking logarithms of both sides in equation (1) and then differentiating, we observe that

$$-\frac{\zeta'}{\zeta}(s) = \sum_{n=1}^{\infty} \frac{\Lambda(n)}{n^s}$$

where

$$\Lambda(n) = \begin{cases} \log p & \text{if } n \text{ is a power of the prime } p \\ 0 & \text{otherwise} \end{cases}$$

denotes the von Mangoldt function (after the mathematician who introduced the notation). From the second equation in (2), we see that $-\zeta'(s)/\zeta(s)$ has a simple pole at $s = 1$ with residue 1. If in addition we knew that $\zeta(s)$ does not vanish on $\operatorname{Re}(s) = 1$, then $-\zeta'(s)/\zeta(s)$ is represented in the region $\operatorname{Re}(s) > 1$, by a Dirichlet series with non-negative coefficients and has a meromorphic continuation to $\operatorname{Re}(s) \geq 1$ with only a simple pole at $s = 1$ with residue 1. Applying the Wiener-Ikehara Tauberian theorem, we deduce that

$$\sum_{n \leq x} \Lambda(n) = x + \mathbf{o}(x)$$

as $x \to \infty$. It is now an easy exercise to deduce the prime number theorem from this asymptotic formula (see exercise 1).

Hadamard's proof that $\zeta(1+it) \neq 0$ for $t \in \mathbb{R}$, is exceedingly simple and can be explained intuitively as follows (see [Ka]). Let us write,

$$\log \zeta(s) = \sum_{n=1}^{\infty} \frac{a_n}{n^s}$$

and note that $a_n \geq 0$. Since $\zeta(s)$ has a simple pole at $s = 1$,

$$\log \zeta(1+\epsilon) = \log 1/\epsilon + \mathbf{O}(1)$$

as $\epsilon \to 0^+$. If $\zeta(s)$ has a zero of order m (say) at $s = 1 + it$, then

$$\log \zeta(1+it+\epsilon) = -m \log 1/\epsilon + \mathbf{O}(1).$$

Therefore, using $a_n \geq 0$, we deduce that $n^{it} \simeq -m$ for most n's. So $m = 1$ and $n^{2it} \simeq 1$ for most n's so that

$$\log \zeta(1+2it+\epsilon) = \log 1/\epsilon$$

which is not possible since $1 + 2it$ is a regular point of $\zeta(s)$ for $t \neq 0$. This proves the non-vanishing.

Though this is an intuitive proof, it captures the essence of the argument. The traditional proof begins by considering for $\sigma > 1$ the combination[2]

$$- \operatorname{Re}\left(3 \log \zeta(\sigma) + 4 \log \zeta(\sigma + it) + \log \zeta(\sigma + 2it)\right)$$
$$= \sum_{n=1}^{\infty} \frac{\Lambda(n)}{n^\sigma \log n} \left(3 + 4\cos(t \log n) + \cos(2t \log n)\right). \tag{3}$$

Since

$$3 + 4\cos\theta + \cos 2\theta = 2(1 + \cos\theta)^2,$$

we see that the right side of equation (3) is non-negative. Hence,

$$|\zeta(\sigma)^3 \zeta(\sigma + it)^4 \zeta(\sigma + 2it)| \geq 1$$

for $t \in \mathbb{R}$ and $\sigma > 1$. If $\zeta(1+it) = 0$ for $t \neq 0$, then as $\sigma \to 1^+$ in the above inequality, the left hand side tends to zero which is a contradiction. This completes the proof that $\zeta(s)$ does not vanish on $\operatorname{Re}(s) = 1$.

We now present a generalization of this non-vanishing result (see [VKM, p. 199]).

---

2)   The traditional proof will use the traditional notation, a curious combination of Greek and Roman letters, in writing $s = \sigma + it$ with σ denoting the real part of s and t the imaginary part. i will of course denote $\sqrt{-1}$.

Theorem 1.2. *Let $f(s)$ be a function satisfying the following hypotheses:*
(a) *f is holomorphic in $\sigma > 1$ and non-zero there;*
(b) *on the line $\sigma = 1$, f is holomorphic except for a pole of order $e \geq 0$ at $s = 1$;*
(c) *$\log f(s)$ can be written as a Dirichlet series $\sum_{n=1}^{\infty} b_n/n^s$ with $b_n \geq 0$, for $\sigma > 1$.*

If f has a zero on the line $\sigma = 1$, then the order of the zero is bounded by $e/2$. (Here we are writing $s = \sigma + it$.)

Proof. Suppose f has a zero at $1 + it_0$ of order $k > e/2$. Then, $e \leq 2k - 1$. Consider the function

$$g(s) = f(s)^{2k+1} \prod_{j=1}^{2k} f(s + ijt_0)^{2(2k+1-j)}$$

$$= f(s)^{2k+1} f(s + it_0)^{4k} f(s + 2it_0)^{4k-2} \cdots f(s + 2kit_0)^2.$$

Then, g is holomorphic for $\sigma > 1$ and vanishes to at least first order at $s = 1$ as

$$4k^2 - (2k+1)e \geq 4k^2 - (2k+1)(2k-1) = 1.$$

However, for $\sigma > 1$,

$$\log g(s) = \sum_{n=1}^{\infty} b_n n^{-s} \left(2k+1 + \sum_{j=1}^{2k} 2(2k+1-j)n^{-ijt_0} \right).$$

Let $\phi_n = t_0 \log n$. Then, for $\sigma > 1$,

$$\mathrm{Re}\,(\log g(\sigma)) = \log |g(\sigma)| = \sum_{n=1}^{\infty} b_n n^{-\sigma} \left(2k+1 + \sum_{j=1}^{2k} 2(2k+1-j)\cos(j\phi_n) \right).$$

Now we have the identity

$$F(k,\theta) := 2k+1 + \sum_{j=1}^{2k} 2(2k+1-j)\cos(j\theta) = \left(1 + 2\sum_{j=1}^{k}\cos j\theta \right)^2 \geq 0$$

(see exercise 2). Hence, $\log |g(\sigma)| \geq 0$ for $\sigma > 1$. That is

$$|g(\sigma)| \geq 1.$$

This contradicts g having a zero at $\sigma = 1$.

By applying the Tauberian theorem to the function $-\zeta'(s)/\zeta(s)$, which by the non-vanishing theorem satisfies the conditions of Theorem 1.1, we deduce the prime number theorem in the form

$$\psi(x) := \sum_{n \leq x} \Lambda(n) \sim x$$

as $x \to \infty$. The famous Riemann hypothesis asserts that $\zeta(s) \neq 0$ for $\mathrm{Re}(s) > 1/2$. If we assume this hypothesis, one can prove by methods of contour integration the formula

$$\psi(x) = x + \mathbf{O}(x^{1/2} \log^2 x).$$

After this brief discussion, we are now ready to prove Theorem 1.1.

Proof of Theorem 1.1. If the a_n are real, it suffices to prove the theorem for $F(s)$, for then one can apply such a result to $F(s) - f(s)$ which is a Dirichlet series with non-negative coefficients in the region $\mathrm{Re}(s) > 1$.

If not all the a_n are real, then set

$$f^*(s) = \sum_{n=1}^{\infty} \overline{a}_n/n^s$$

and observe that

$$f = \frac{1}{2}(f + f^*) + \frac{i}{2}\left(\frac{f - f^*}{i}\right).$$

Then it suffices to prove the theorem for $F(s)$. We begin by reviewing certain facts from Fourier analysis (see [Rudin]).

Let

$$S = \{f \in C^\infty(\mathbb{R}) \mid \text{for all } n, m \in \mathbb{Z}^+, \lim_{|x| \to \infty} x^n \frac{d^m}{dx^m} f(x) = 0\}.$$

For functions $f \in S$, we have the *Fourier inversion*:

$$\hat{f}(x) = \frac{1}{\sqrt{2\pi}} \int_{-\infty}^{\infty} f(t) e^{-itx} \, dt$$

$$f(x) = \frac{1}{\sqrt{2\pi}} \int_{-\infty}^{\infty} \hat{f}(t) e^{itx} \, dt.$$

Hence,

$$\hat{f}(x - y) = \frac{1}{\sqrt{2\pi}} \int_{-\infty}^{\infty} f(t) e^{ity} e^{-itx} \, dt$$

so that $\hat{f}(x - y)$ and $f(t) e^{ity}$ are transforms of each other. The formula

$$\int_{-\infty}^{\infty} f(x) g(x) \, dx = \int_{-\infty}^{\infty} \hat{f}(t) \hat{g}(t) \, dt$$

is known as *Parseval's formula*. The *Riemann-Lebesgue lemma* asserts that

$$\lim_{\lambda \to \infty} \int_{-\infty}^{\infty} f(t) \, e^{i\lambda t} \, dt = 0,$$

for absolutely integrable functions. The *Féjer Kernel*

$$K_\lambda(x) = \frac{\sin^2 \lambda x}{\lambda x^2}$$

has Fourier transform

$$\hat{K}_\lambda(x) = \begin{cases} 2\sqrt{2}\,\pi \left(1 - \frac{|x|}{2\lambda}\right) & \text{if } |x| \leq 2\lambda \\ 0 & \text{otherwise.} \end{cases}$$

Now let $F(s) = \sum_{n=1}^{\infty} b_n/n^s$, $b_n \geq 0$ be as above and define an analytic function for $\mathrm{Re}(s) > 1$. Put $B(x) = \sum_{n \leq x} b_n$. Replacing b_n by b_n/R, we can suppose without loss of generality that $R = 1$. Then, by partial summation, we see that for $\mathrm{Re}(s) > 1$,

$$F(s) = s \int_1^{\infty} \frac{B(x)}{x^{s+1}} \, dx.$$

Set $x = e^u$. Then

$$\frac{F(s)}{s} = \int_0^{\infty} B(e^u) \, e^{-us} \, du.$$

Note that

$$\int_0^{\infty} e^{-u(s-1)} \, du = \frac{1}{s-1}.$$

Hence, putting $s = 1 + \delta + it$, $\delta > 0$, we get

$$\frac{F(1+\delta+it)}{1+\delta+it} - \frac{1}{s-1} = \int_0^{\infty} \left(B(e^u) \, e^{-u} - 1\right) e^{-u\delta} \, e^{-iut} \, du.$$

Set

$$g(u) = B(e^u) e^{-u}, \qquad h_\delta(t) = \frac{F(1+\delta+it)}{1+\delta+it} - \frac{1}{s-1},$$

and

$$h(t) = \frac{F(1+it)}{1+it} - \frac{1}{s-1} \qquad (s = 1+it),$$

which is regular for t in $\mathbb{R}$.

Our goal is to prove that $g(u) \to 1$ as $u \to \infty$. The above formula says that the Fourier transform of $\sqrt{2\pi}(g(u) - 1)e^{-u\delta}$ is $h_\delta(t)$. Applying Parseval's formula, we deduce

$$\int_{-\infty}^{\infty} (g(u) - 1)e^{-u\delta} K_\lambda(u) \, du = \int_{-\infty}^{\infty} h_\delta(t)\hat{K}_\lambda(t) \, dt.$$

But note that we also have by Parseval's formula

$$\int_{-\infty}^{\infty} (g(u) - 1)e^{-u\delta} K_\lambda(u - v)\, du = \int_{-\infty}^{\infty} h_\delta(t)\hat{K}_\lambda(t)e^{itv}\, dt.$$

Since $\hat{K}_\lambda$ has compact support, the limit as $\delta \to 0$ of the right hand side exists. The same is true of the left hand side. Hence

$$\int_{-\infty}^{\infty} (g(u) - 1))K_\lambda(u - v)\, du = \int_{-\infty}^{\infty} h(t)\hat{K}_\lambda(t)e^{itv}\, dt.$$

By the Riemann-Lebesgue lemma, we deduce

$$\lim_{v \to \infty} \int_{-\infty}^{\infty} (g(u) - 1))K_\lambda(u - v)\, du = 0.$$

Thus,

$$\lim_{v \to \infty} \int_{-\infty}^{\infty} g(u)K_\lambda(u - v)\, du = \pi.$$

Set $-\lambda(u - v) = \alpha$. Then $u = v - \frac{\alpha}{\lambda}$ and so as g is bounded,

$$\lim_{v \to \infty} \int_{-\infty}^{v\lambda} g\left(v - \frac{\alpha}{\lambda}\right) \frac{\sin^2 \alpha}{\alpha^2}\, d\alpha = \pi.$$

We can now prove the theorem. Since $B(x)$ is monotone increasing, we see that

$$g(u_2) \geq g(u_1)e^{u_1 - u_2}, \qquad u_1 \leq u_2.$$

Thus, for $|\alpha| \leq \sqrt{\lambda}$, we have

$$g\left(v - \frac{\alpha}{\lambda}\right) \geq g\left(v - \frac{1}{\sqrt{\lambda}}\right) e^{-\frac{1}{\sqrt{\lambda}} + \frac{\alpha}{\lambda}} \geq g\left(v - \frac{1}{\sqrt{\lambda}}\right) e^{-\frac{2}{\sqrt{\lambda}}}.$$

Since

$$\limsup_{v \to \infty} \int_{-\sqrt{\lambda}}^{\sqrt{\lambda}} g\left(v - \frac{\alpha}{\lambda}\right) \frac{\sin^2 \alpha}{\alpha^2}\, d\alpha \leq \pi,$$

we deduce,

$$\limsup_{v \to \infty} g\left(v - \frac{1}{\sqrt{\lambda}}\right) e^{-\frac{2}{\sqrt{\lambda}}} \leq \frac{\pi}{\int_{-\sqrt{\lambda}}^{\sqrt{\lambda}} \frac{\sin^2 \alpha}{\alpha^2}\, d\alpha}.$$

Since v is arbitrary, changing v to $v + \frac{1}{\sqrt{\lambda}}$, we get

$$\limsup_{v \to \infty} g(v) \leq 1.$$

The lower bound is obtained similarly:

$$\liminf_{v\to\infty} \int_{-\sqrt{\lambda}}^{\sqrt{\lambda}} g\left(v - \frac{\alpha}{\lambda}\right) \frac{\sin^2 \alpha}{\alpha^2} d\alpha \geq \pi + \mathbf{O}\left(\frac{1}{\sqrt{\lambda}}\right).$$

Since

$$g\left(v - \frac{\alpha}{\lambda}\right) \leq g\left(v + \frac{1}{\sqrt{\lambda}}\right) e^{v + \frac{1}{\sqrt{\lambda}} - v + \frac{\alpha}{\lambda}} \leq g\left(v + \frac{1}{\sqrt{\lambda}}\right) e^{\frac{2}{\sqrt{\lambda}}},$$

we obtain

$$\liminf_{v\to\infty} g\left(v + \frac{1}{\sqrt{\lambda}}\right) e^{\frac{2}{\sqrt{\lambda}}} \int_{-\sqrt{\lambda}}^{\sqrt{\lambda}} \frac{\sin^2 \alpha}{\alpha^2} d\alpha \geq \pi + \mathbf{O}\left(\frac{1}{\sqrt{\lambda}}\right),$$

so that

$$\liminf_{v\to\infty} g(v) \geq 1,$$

as desired. Together with

$$\limsup_{v\to\infty} g(v) \leq 1,$$

we deduce $\lim_{v\to\infty} g(v) = 1$ as needed. This completes the proof of the theorem.

By the same method, one can deduce the following variation:

Theorem 1.3. *Suppose that the function $F(s)$ has the following properties:*
(a) *there exists $\beta > 0$ such that for $\mathrm{Re}(s) > \beta$,*

$$F(s) = s \int_0^\infty B(u) e^{-us} du$$

where $B(u)$ is a positive monotone increasing function;
(b) *there exist constants $\alpha > -1$, $c > 0$ such that*

$$F(s) = \frac{H(s)}{(s - \beta)^{\alpha+1}} \text{ and } H(\beta) = c\Gamma(\alpha + 1)$$

where H is holomorphic in $\mathrm{Re}(s) \geq \beta$. Then

$$B(u) = (c + \mathbf{o}(1)) u^\alpha e^{\beta u}$$

as $u \to \infty$.

Corollary 1.4. *Suppose that for* Re$(s) > \beta$,

$$F(s) = \sum_{n=1}^{\infty} b_n/n^s$$

with $b_n \geq 0$, *and that for* Re$(s) \geq \beta$, $F(s)$ *admits a meromorphic continuation with at most a pole of order* $\alpha + 1$ $(\alpha > -1)$ *at* $s = \beta$. *Then*

$$\sum_{n \leq x} b_n = (c + \mathbf{o}(1))x^{\beta} \log^{\alpha} x$$

as $x \to \infty$.

This formulation is useful in most applications. There are other variations (e.g. see Ellison [p. 64–65]) where upper and lower bounds for $\sum_{n \leq x} b_n$ can be obtained from the knowledge of analytic continuation for Re$(s) \geq \beta$, $|\text{Im}(s)| \leq T$.

We can apply this to the Riemann zeta function. Indeed,

$$\zeta(s) = s \int_{1}^{\infty} \frac{[x]}{x^{s+1}}\, dx = \frac{s}{s-1} - s \int_{1}^{\infty} \frac{\{x\}}{x^{s+1}}\, dx$$

gives a meromorphic continuation of $\zeta(s)$ for Re$(s) > 0$ with only a simple pole at $s = 1$ and residue 1. If $d_k(n)$ denotes the number of ways of writing n as a product of k positive numbers, it is easily seen that

$$\zeta^k(s) = \sum_{n=1}^{\infty} \frac{d_k(n)}{n^s}.$$

Hence, by Corollary 1.4 we deduce that

$$\sum_{n \leq x} d_k(n) = (1 + \mathbf{o}(1))\frac{x \log^{k-1} x}{\Gamma(k)}$$

as $x \to \infty$.

§2 Primes in Arithmetic Progression

In 1837, Dirichlet proved the infinitude of primes in a given arithmetic progression. Historically, this work preceded Riemann's paper on the zeta function. Dirichlet proved his theorem by introducing the L-functions $L(s, \chi)$ which now bear his name. However, he treated them as functions of a real variable only and therefore obtained results of the form

$$\lim_{s \to 1+} \sum_{p \equiv a (\bmod q)} \frac{1}{p^s} = +\infty,$$

where the summation is over primes $p \equiv a \pmod q$. He first proved his theorem for prime modulus q and then a year later, treated the general case. In the course of this discovery, he contributed two fundamental ideas: (a) the beginnings of the theory of group characters and (b) the celebrated class number formula. The first was essential in 'sifting' the primes in a given arithmetic progression. The second was used to establish the non-vanishing of certain of his L-functions needed in his proof.

If $(\mathbb{Z}/q\mathbb{Z})^*$ is the multiplicative group of coprime residue classes, let

$$\chi : (\mathbb{Z}/q\mathbb{Z})^* \to \mathbb{C}^*$$

be a homomorphism. (These are called Dirichlet characters.) One now defines for any $n \in \mathbb{Z}$,

$$\chi(n) = \begin{cases} \chi(n \bmod q) & \text{if } (n, q) = 1 \\ 0 & \text{otherwise} \end{cases}$$

and (by abuse of language) we also call these Dirichlet characters. There are $\phi(q)$ such characters where ϕ is Euler's function and we denote by χ_0 the trivial character. Analogous to the Riemann zeta function, we define

$$L(s, \chi) = \sum_{n=1}^{\infty} \frac{\chi(n)}{n^s}.$$

Since $|\chi(n)| \le 1$ for all values of n, the series converges absolutely for $\mathrm{Re}(s) > 1$. If we write

$$S(x) = \sum_{n \le x} \chi(n),$$

then as in section one, we can write

$$L(s, \chi) = s \int_1^{\infty} \frac{S(x)}{x^{s+1}} dx. \tag{4}$$

If $\chi \ne \chi_0$, then

$$\sum_{n \bmod q} \chi(n) = 0$$

so that we easily see $|S(x)| \le q$ upon partitioning the interval $[1, x]$ into subintervals equal to or at most of length q. Hence, (4) converges for $\mathrm{Re}(s) > 0$ and this gives us an analytic continuation for $L(s, \chi)$ when $\chi \ne \chi_0$, in this half-plane.

As in the case of the Riemann zeta function, the multiplicativity of $\chi(n)$ and the unique factorization of the natural numbers combine to give the Euler product:

$$L(s, \chi) = \prod_p \left(1 - \frac{\chi(p)}{p^s}\right)^{-1}.$$

Notice again, that this product shows $L(s, \chi) \neq 0$ for $\mathrm{Re}(s) > 1$. Taking logarithms and using the orthogonality relations of the characters, we obtain

$$\sum_{p^k \equiv a \bmod q} \frac{1}{kp^{ks}} = \frac{1}{\phi(q)} \sum_{\chi \bmod q} \bar{\chi}(a) \log L(s, \chi)$$

valid for $\mathrm{Re}(s) > 1$. Dirichlet noticed that

$$L(s, \chi_0) = \left[\prod_{p | q} \left(1 - \frac{1}{p^s} \right) \right] \zeta(s)$$

and hence

$$\lim_{s \to 1^+} \log L(s, \chi_0) = +\infty$$

by the divergence of the harmonic series. For $\chi \neq \chi_0$, we know $L(s, \chi)$ is regular for $\mathrm{Re}(s) > 0$. If $L(1, \chi) \neq 0$ for all $\chi \neq \chi_0$, then we deduce that

$$\lim_{s \to 1^+} \sum_{p^k \equiv a \bmod q} \frac{1}{kp^{ks}} = +\infty$$

from which we infer

$$\lim_{s \to 1^+} \sum_{p \equiv a \bmod q} \frac{1}{p^s} = +\infty,$$

since the contribution for $k \geq 2$ in the penultimate sum remains bounded as $s \to 1^+$.

We will now establish the non-vanishing of $L(s, \chi)$ on $\mathrm{Re}(s) = 1$ by appealing to Theorem 1.2. But before we do, we need the following lemma, which is a variation on a well-known result of Landau (see exercise 5).

Lemma 2.1.   Let $g(s) = \sum_{n=1}^{\infty} a_n / n^s$ be a Dirichlet series where $a_1 = 1$ and $a_n \geq 0$. Suppose that the series is absolutely convergent in $\mathrm{Re}(s) > 1$. Suppose further that $g(s)$ admits an analytic continuation to $\mathrm{Re}(s) \geq 1/2$. Then $g(1/2) \neq 0$.

Proof. We can expand $g(s)$ as a Laurent series about the point $s = 2$ in a disc $|s - 2| < 3/2$:

$$g(s) = \sum_{m=0}^{\infty} \frac{g^{(m)}(2)}{m!} (s - 2)^m.$$

Computing $g^{(m)}(2)$ explicitly from the Dirichlet series

$$g^{(m)}(2) = (-1)^m \sum_{n=1}^{\infty} a_n (\log n)^m n^{-2} = (-1)^m b_m \qquad \text{(say)},$$

where $b_m \geq 0$. Hence,

$$g(s) = \sum_{m=0}^{\infty} \frac{b_m}{m!} (2-s)^m$$

valid for $|2-s| < 3/2$. In particular, if $1/2 < s < 2$, then

$$g(s) \geq g(2) \geq 1$$

since all the terms are non-negative. Taking $\lim_{s \to 1/2+} g(s)$ gives $g(1/2) \geq 1$. This establishes the result.

Theorem 2.2. $L(1+it, \chi) \neq 0$ *for all $t \in \mathbb{R}$ and all $\chi \neq \chi_0$.*

Proof. We will apply Theorem 1.2 to the function

$$f(s) = \prod_{\chi} L(s, \chi).$$

Observe first that

$$\log f(s) = \sum_{\chi} \log L(s, \chi) = \sum_{p^k \equiv 1 \bmod q} \frac{1}{k p^{ks}}$$

is a Dirichlet series with non-negative coefficients absolutely convergent in $\mathrm{Re}(s) > 1$. Since $L(s, \chi_0)$ has a simple pole at $s = 1$ and $L(s, \chi)$ is regular in $\mathrm{Re}(s) > 0$ for $\chi \neq \chi_0$, we find that $f(s)$ satisfies (a), (b), (c) of the theorem with $e \leq 1$. Hence any zero of $f(s)$ is of order bounded by $1/2$.

Therefore $L(1+it, \chi) \neq 0$ for $t \in \mathbb{R}, t \neq 0$. We must still analyse the possibility $L(1, \chi) = 0$. If $L(1, \chi) = 0$ for some non-real character χ, then $L(1, \overline{\chi}) = 0$ and $f(s)$ would have a zero at $s = 1$ contrary to what has already been established. So we need to look at the possibility $L(1, \chi) = 0$ for a real-valued character. If this were the case, we consider

$$g(s) = \frac{L(s, \chi_0) L(s, \chi)}{L(2s, \chi_0)} = \prod_p \left(1 + \frac{\chi_0(p)}{p^s}\right) \left(1 - \frac{\chi(p)}{p^s}\right)^{-1}$$

which is a Dirichlet series with non-negative coefficients because the Euler product is supported only at primes where $\chi(p) = +1$. Note that if $L(1, \chi) = 0$, then this cancels the pole of $L(s, \chi_0)$ at $s = 1$ so that $g(s)$ is regular there. In fact, $g(s)$ is regular for $\mathrm{Re}(s) \geq 1/2$ since the numerator is regular for $\mathrm{Re}(s) \geq 0$ and the denominator does not vanish for $\mathrm{Re}(s) \geq 1/2$ essentially by Theorem 1.1. Moreover, since $L(2s, \chi_0)$ has a simple pole at $s = 1/2$, $g(s)$ has a zero at $s = 1/2$ which contradicts Lemma 2.1. This completes the proof.

Applying the Tauberian theorem to

$$- \sum_{\chi \bmod q} \overline{\chi}(a) \frac{L'}{L}(s, \chi)$$

we immediately deduce that

$$\psi(x, q, a) := \sum_{\substack{n \leq x \\ n \equiv a \bmod q}} \Lambda(n) \sim \frac{x}{\phi(q)},$$

which is the prime number theorem for arithmetic progressions.

§3 Dedekind's zeta function

Let K be an algebraic number field of finite degree n over $\mathbb{Q}$. The zeta function of K is defined as

$$\zeta_K(s) = \sum_{\mathfrak{a}} (\mathbb{N}\mathfrak{a})^{-s}$$

where the sum is over all integral ideals of $\mathcal{O}_K$, the ring of integers of K. Because of Dedekind's theorem that every ideal of $\mathcal{O}_K$ can be factored as a product of prime ideals uniquely, we have the Euler product formula:

$$\zeta_K(s) = \prod_{\mathfrak{p}} \left(1 - \frac{1}{\mathbb{N}\mathfrak{p}^s}\right)^{-1},$$

where the product is over all prime ideals of K. Notice that this product shows $\zeta_K(s) \neq 0$ for $\mathrm{Re}(s) > 1$. Hecke proved in 1917, that $(s-1)\zeta_K(s)$ extends to an entire function such that

$$\lim_{s \to 1^+} (s-1)\zeta_K(s) = \kappa = \frac{2^{r_1}(2\pi)^{r_2} h R}{w\sqrt{|d_K|}}$$

where r_1 is the number of real conjugate fields, $2r_2$ is the number of complex conjugate fields, h is the class number, R is the regulator, w is the number of roots of unity and d_K is the discriminant of K. Moreover, $\zeta_K(s)$ satisfies a functional equation

$$\xi_K(s) = \xi_K(1-s)$$

where

$$\xi_K(s) = \left(\frac{\sqrt{|d_K|}}{2^{r_2}\pi^{n/2}}\right)^s \Gamma(s/2)^{r_1}\Gamma(s)^{r_2}\zeta_K(s).$$

The functional equation allows one to write the residue in a slightly elegant form

$$\lim_{s \to 0} \frac{\zeta_K(s)}{s^r} = -\frac{hR}{w}$$

where $r = r_1 + r_2 - 1$. By the Tauberian Theorem 1.1, we immediately deduce

Theorem 3.1. *Let a_m be the number of ideals of K of norm m. Then,*

$$\sum_{m \leq x} a_m \sim \kappa x$$

as $x \to \infty$.

Proof. Since $\zeta_K(s)$ has a simple pole at $s = 1$, the result follows from Theorem 1.1.

As before, we can consider

$$-\frac{\zeta_K'}{\zeta_K}(s) = \sum_{\mathfrak{a}} \frac{\Lambda(\mathfrak{a})}{N\mathfrak{a}^s}$$

where

$$\Lambda(\mathfrak{a}) = \begin{cases} \log N\mathfrak{p} & \text{if } \mathfrak{a} = \mathfrak{p}^m \text{ for some prime ideal } \mathfrak{p} \\ 0 & \text{otherwise,} \end{cases}$$

is the number field analogue of the von Mangoldt function. It is now clear that we can apply the Tauberian theorem to $-\zeta_K'(s)/\zeta_K(s)$ to deduce the prime ideal theorem:

Theorem 3.2. *Let $\pi_K(x)$ denote the number of prime ideals of K whose norm is $\leq x$. Then*

$$\pi_K(x) \sim \frac{x}{\log x},$$

as $x \to \infty$.

Proof. For $\mathrm{Re}(s) > 1$, $\log \zeta_K(s)$ is a Dirichlet series with non-negative coefficients by virtue of the Euler product. Moreover, $\zeta_K(s)$ is holomorphic on $\mathrm{Re}(s) = 1$ except at $s = 1$ where it has a simple pole. Hence, we can apply Theorem 1.2 to deduce $\zeta_K(1 + it) \neq 0$ for all $t \in \mathbb{R}$. Applying Theorem 1.1 we deduce

$$\sum_{N\mathfrak{a} \leq x} \Lambda(\mathfrak{a}) \sim x$$

as $x \to \infty$. We now deduce the result by partial summation (see exercise 1).

Let us consider the special case $K = \mathbb{Q}(i)$. If $r(n)$ denotes the number of ways of writing n as a sum of two integer squares, then Theorem 3.1 gives

$$\sum_{n \leq x} r(n) \sim \pi x$$

as $x \to \infty$, since $\mathbb{Z}[i]$ is a unique factorization domain and thus has class number 1. It is also not difficult to see that a rational prime p is the norm of a prime ideal $\mathbb{Z}[i]$ if and only if p can be written as the sum of two integral squares. If p can be so written, and p is odd, there are exactly two prime ideals of $\mathbb{Z}[i]$ of norm p. Thus, Theorem 3.2 in this case proves that the number of primes $p \leq x$ which can be written as the sum of two squares is

$$\sim \frac{1}{2} \frac{x}{\log x}$$

as $x \to \infty$.

§4 Hecke's *L*-functions

We begin by constructing the analogues of Dirichlet's *L*-functions. We first need to define the notion of "ideal classes" and then define characters of these classes.

Let K be an algebraic number field and $\mathfrak{f}$ an ideal of $\mathcal{O}_K$. A natural starting point is to consider the ideal class group and to define characters of this group. One can generalize this to obtain a notion of ideal classes mod $\mathfrak{f}$ as follows. The multiplicative group generated by all ideals coprime to $\mathfrak{f}$ will be denoted by $I(\mathfrak{f})$. The principal ray class $P(\mathfrak{f})$ (mod $\mathfrak{f}$) is the subgroup of principal ideals of the form (α/β) with

    (i) $\alpha, \beta \in \mathcal{O}_K$ and coprime to $\mathfrak{f}$;

    (ii) $\alpha \equiv \beta \pmod{\mathfrak{f}}$;

    (iii) α/β is totally positive (that is, all its real conjugates are positive).

The quotient group $G(\mathfrak{f}) = I(\mathfrak{f})/P(\mathfrak{f})$ is called the ray class group mod $\mathfrak{f}$. The elements of this group are called ray classes. These will be considered as analogues of the groups $(\mathbb{Z}/m\mathbb{Z})^*$ in the rational number field case. Let us note that without the totally positive condition, the construction leads to $(\mathbb{Z}/m\mathbb{Z})^*/\{\pm 1\}$ if $K = \mathbb{Q}$ and $\mathfrak{f} = (m)$.

Let χ be a character of the abelian group $G(\mathfrak{f})$. Define

$$L(s, \chi) = \sum_{\mathfrak{a}} \frac{\chi(\mathfrak{a})}{N\mathfrak{a}^s}$$

where the sum is over integral ideals $\mathfrak{a}$ of K. This series converges absolutely for $\mathrm{Re}(s) > 1$ as is seen by comparing with the Dedekind zeta function which converges absolutely in that region. We again have the Euler product:

$$L(s, \chi) = \prod_{\mathfrak{p}} \left(1 - \frac{\chi(\mathfrak{p})}{N\mathfrak{p}^s}\right)^{-1}$$

which is valid for $\mathrm{Re}(s) > 1$. This product shows the non-vanishing of $L(s, \chi)$ in that region. If $\chi \neq \chi_0$, the trivial character, Hecke showed that $L(s, \chi)$ extends to an entire function.

By considering

$$f(s) = \zeta_K(s)L(s, \chi)L(s, \overline{\chi})L(s, \chi\overline{\chi})$$

and applying Theorem 1.2, we deduce that $f(s)$ does not vanish on $\mathrm{Re}(s) = 1$ for $s \neq 1$. In the latter case, we consider

$$g(s) = \zeta_K(s)^3 L(s, \chi)^4 L(s, \chi^2)$$

which has non-negative coefficients. We can apply Theorem 1.2 again to get the non-vanishing of $L(1, \chi)$ provided $\chi^2 \neq 1$. In the last case, we need to consider as before

$$h(s) = \frac{\zeta_K(s)L(s, \chi)}{\zeta_K(2s)}$$

and apply the reasoning as before. This allows us to deduce

Theorem 4.1.   *The number of prime ideals with norm less than x, in a given ray class is*

$$\frac{1}{|G(\mathfrak{f})|}\frac{x}{\log x}$$

as $x\to\infty$.

We can consider a more general situation. Recall that each finite prime $\mathfrak{p}$ defines a valuation $v_\mathfrak{p} : K\to\mathbb{Z}$ given by $v_\mathfrak{p}(\alpha) = $ the exponent of $\mathfrak{p}$ in the prime ideal decomposition of the principal ideal (α). We can extend this definition to ideals. A generalized ideal $\mathfrak{a}$ of K is an ideal $\mathfrak{a}_f$ together with a set of embeddings $\{\sigma_1,\ldots,\sigma_i\}$ of K into $\mathbb{R}$. We will say that $\alpha \in K$ satisfies the congruence $\alpha \equiv 1(\mathrm{mod}\,\mathfrak{a})$ if α is a unit at all the primes dividing $\mathfrak{a}_f$, $v_\mathfrak{p}(\alpha - 1) \geq v_\mathfrak{p}(\mathfrak{a}_f)$ and $\sigma_j(\alpha) > 0$ for $j = 1,\ldots,i$. We can now define $G(\mathfrak{a})$ as the quotient group of fractional ideals coprime to $\mathfrak{a}_f$ modulo principal ideals (α) with $\alpha \equiv 1(\mathrm{mod}\,\mathfrak{a})$. Hecke defined L-series for characters of these generalized ideal class groups and an analogue of Theorem 4.1 is true for prime ideals in a given ideal class.

We recomend that the reader study the theory of Hecke L-functions as explained, for example, in Lang [L], both from the classical and adelic points of view (Tate's thesis).

Exercises

1. Let $f(t)$ have a continuous derivative $f'(t)$, for $t \geq 1$. Let c_n for $n \geq 1$ be constants and let

$$C(u) = \sum_{n\leq u} c_n.$$

Then, prove that

$$\sum_{n\leq x} c_n f(n) = f(x)C(x) - \int_1^x f'(t)C(t)dt,$$

and that

$$\sum_{n\leq x} f(n) = \int_1^x f(t)dt + \int_1^x (t - [t])f'(t)dt + f(1) - (x - [x])f(x).$$

Deduce that

$$\sum_{n\leq x} \Lambda(n) = x + \mathbf{o}(x)$$

as $x\to\infty$ if and only if

$$\lim_{x\to\infty} \frac{\pi(x)}{x/\log x} = 1.$$

2. Prove that

$$2k + 1 + \sum_{j=1}^{2k} 2(2k + 1 - j)\cos(j\theta) = \left(1 + 2\sum_{j=1}^{k}\cos j\theta\right)^2.$$

Notice that the case $k = 1$ is the classical trigonometric identity of Hadamard and de la Vallée Poussin.

3. Suppose we had a trigonometric polynomial

$$a_0 + a_1\cos\theta + \cdots a_n\cos n\theta \geq 0.$$

Show that in the proof of Theorem 1.2, we obtain

$$k \leq |a_0/a_1|e.$$

Show further that

$$|a_0/a_1| > 1/2.$$

(See Féjer [Fe].)

4. For $(a, b) = 1$, compute gcd $\{\phi(an + b) : n \in \mathbb{Z}\}$. What about the same question for $\phi(an^2 + bn + c)$?

5. (Landau's theorem) Suppose $a_n \geq 0$ and

$$f(s) = \sum_{n=1}^{\infty} \frac{a_n}{n^s}$$

has abscissa of convergence equal to α. Show that $f(s)$ has a singularity at $s = \alpha$.

References

[Ellison] W. Ellison, Prime numbers, John Wiley and Sons, Paris, Hermann, 1985.

[Fe] L. Féjer, Über trigonometrische Polynome, *J. Reine Angew. Math.*, **146** (1916), pp. 53–82.

[Ka] Jean-Pierre Kahane, Jacques Hadamard, *Math. Intelligencer*, Vol. 13, No. 1, (1991), pp. 23–29.

[L] S. Lang, Algebraic Number Theory, Springer-Verlag, 1986.

[VKM] V. Kumar Murty, On the Sato-Tate conjecture, in Number Theory related to Fermat's Last Theorem, (ed. N. Koblitz), *Progress in Mathematics*, Vol. 26, (1982), pp. 195–205.

[Rudin] W. Rudin, Real and Complex analysis, Bombay, Tata Mcgraw-Hill Publishing Co. Ltd., 1976.

Chapter 2
Artin L-Functions

§1 Group-theoretic background

In this section, we shall collect together a few group theoretic preliminaries. We begin by reviewing the basic aspects of characters and class functions.

Let G be a finite group. If $f_1, f_2 : G \to \mathbb{C}$ are two $\mathbb{C}$-valued functions on G, we define their inner product by

$$(f_1, f_2) = \frac{1}{|G|} \sum_{g \in G} f_1(g) \overline{f_2(g)}.$$

If $f : G \to \mathbb{C}$ is a $\mathbb{C}$-valued function on G, and $\sigma \in G$, we define $f^\sigma : G \to \mathbb{C}$ by $f^\sigma(g) = f(\sigma g \sigma^{-1})$. We say f is a *class function* if $f^\sigma = f$ for all $\sigma \in G$.

Let $H \subseteq G$ be a subgroup and $f : H \to \mathbb{C}$ a class function on H. We define a class function

$$\mathrm{Ind}_H^G f : G \to \mathbb{C}$$

on G as follows. Let $g_1, \ldots, g_r$ ($r = [G : H]$) be coset representatives for H in G (so that $G = \cup g_i H$). Extend f to a function $\dot{f}$ on G by setting

$$\dot{f}(g) = \begin{cases} f(g) & g \in H \\ 0 & g \notin H \end{cases}$$

Then

$$(\mathrm{Ind}_H^G f)(g) = \sum_{i=1}^{r} \dot{f}(g_i^{-1} g g_i) = \frac{1}{|H|} \sum_{s \in G} \dot{f}(s^{-1} g s).$$

Let f_1 be a class function on the subgroup H and f_2 a class function on G. The *Frobenius reciprocity theorem* tells us that

$$(f_1, f_2 \,|_H) = (\mathrm{Ind}_H^G f_1, f_2).$$

25

Let H_1, H_2 be subgroups of G and let f be a class function on H_2. Suppose that $G = H_1 H_2$. Then one of *Mackey's theorems* tells us that

$$(\mathrm{Ind}_{H_2}^G f)|_{H_1} = \mathrm{Ind}_{H_1 \cap H_2}^{H_1}(f|_{H_1 \cap H_2}).$$

Let $\rho : G \to GL_n(\mathbb{C})$ be an irreducible representation of G and set $\chi = Tr\,\rho$, the character of ρ. Then χ is a class function on G and every class function is a $\mathbb{C}$-linear combination of characters χ of irreducible representations. A class function which is a $\mathbb{Z}$-linear combination of characters will be called a *generalized character*.

For each $g \in G$, define a symbol x_g and consider the $\mathbb{C}$-vector space

$$V = \oplus_{g \in G} \mathbb{C} x_g.$$

If $|G| = n$, then $\dim V = n$. The *regular representation* reg_G of G

$$\mathrm{reg}_G : G \to GL(V)$$

is defined by

$$\sigma \mapsto (x_g \mapsto x_{\sigma g}).$$

Its character will be denoted by the same letter and we easily see that

$$\mathrm{reg}_G(\sigma) = \begin{cases} n & \sigma = e \text{ (identity)} \\ 0 & \sigma \neq e. \end{cases}$$

In terms of characters

$$\mathrm{reg}_G = \sum_\chi \chi(1)\chi$$

where the sum is over all irreducible characters of G. In terms of induction,

$$\mathrm{reg}_G = \mathrm{Ind}_{\{e\}}^G 1$$

where 1 denotes the (trivial) character of the identity subgroup $\{e\}$.

The reader is referred to Serre [Se1] for an excellent introduction to the representation theory of finite groups.

§2 Definition and basic properties of Artin L-functions

Now let L/K be a Galois extension of number fields, with group G. For each prime $\mathfrak{p}$ of K, and a prime $\mathfrak{q}$ of L with $\mathfrak{q}|\mathfrak{p}$, we define the *decomposition group* $D_{\mathfrak{q}}$ to be $\mathrm{Gal}(L_{\mathfrak{q}}/K_{\mathfrak{p}})$ where $L_{\mathfrak{q}}$ (resp. $K_{\mathfrak{p}}$) is the completion of L (resp. K) at $\mathfrak{q}$ (resp. $\mathfrak{p}$). We have a map from $D_{\mathfrak{q}}$ to $\mathrm{Gal}(k_{\mathfrak{q}}/k_{\mathfrak{p}})$ (the Galois group of the residue field extension) which by Hensel's lemma is surjective. The kernel $I_{\mathfrak{q}}$ is the *inertia group*. We thus have an exact sequence

$$1 \to I_{\mathfrak{q}} \to D_{\mathfrak{q}} \to \mathrm{Gal}(k_{\mathfrak{q}}/k_{\mathfrak{p}}) \to 1.$$

The group $\mathrm{Gal}(k_{\mathfrak{q}}/k_{\mathfrak{p}})$ is cyclic with a generator $x \mapsto x^{\mathbf{N}\mathfrak{p}}$ where $\mathbf{N}\mathfrak{p}$ is the cardinality of $k_{\mathfrak{p}}$. We can *choose* an element $\sigma_{\mathfrak{q}} \in D_{\mathfrak{q}}$ whose image in $\mathrm{Gal}(k_{\mathfrak{q}}/k_{\mathfrak{p}})$ is this generator. We call $\sigma_{\mathfrak{q}}$ a *Frobenius element* at $\mathfrak{q}$ and it is only defined mod $I_{\mathfrak{q}}$. We have $I_{\mathfrak{q}} = 1$ for all unramified $\mathfrak{p}$ (and in particular, these are all but finitely many $\mathfrak{p}$) and so for these $\mathfrak{p}$, $\sigma_{\mathfrak{q}}$ is well-defined. If we choose another prime $\mathfrak{q}'$ above $\mathfrak{p}$,then $I_{\mathfrak{q}'}$ and $D_{\mathfrak{q}'}$ are conjugates of $I_{\mathfrak{q}}$ and $D_{\mathfrak{q}}$. For $\mathfrak{p}$ unramified, we denote by $\sigma_{\mathfrak{p}}$ the *conjugacy class* of Frobenius elements at primes $\mathfrak{q}$ above $\mathfrak{p}$.

Let ρ be a representation of G :

$$\rho : G \to \mathrm{GL}_n(\mathbb{C}).$$

Let χ denote its character. For $\mathrm{Re}(s) > 1$, we define the partial L-function by

$$L_{\mathrm{unramified}}(s, \chi, K) = \prod_{\mathfrak{p} \text{ unramified}} \det(I - \rho(\sigma_{\mathfrak{p}})(\mathbf{N}\mathfrak{p})^{-s})^{-1}$$

where the product is over primes $\mathfrak{p}$ of K with $I_{\mathfrak{q}} = 1$ for any $\mathfrak{q}$ of L with $\mathfrak{q}|\mathfrak{p}$. To obtain an L-function which has good analytic properties (such as functional equation), it is necessary to also define Euler factors at the primes $\mathfrak{p}$ which are ramified in L and also at infinite primes of K.

Let $\mathfrak{p}$ be a prime of K which is ramified in L, and $\mathfrak{q}$ a prime of L above $\mathfrak{p}$. Let V be the underlying complex vector space on which ρ acts. Then we may restrict this action to the decomposition group $D_{\mathfrak{q}}$ and we see that the quotient $D_{\mathfrak{q}}/I_{\mathfrak{q}}$ acts on the subspace $V^{I_{\mathfrak{q}}}$ of V on which $I_{\mathfrak{q}}$ acts trivially. Now we see that any $\sigma_{\mathfrak{q}}$ will have the same characteristic polynomial on this subspace and we define the Euler factor at $\mathfrak{p}$ to be this polynomial:

$$L_{\mathfrak{p}}(s, \chi, K) = \det(I - \rho(\sigma_{\mathfrak{q}})|V^{I_{\mathfrak{q}}}(\mathbf{N}\mathfrak{p})^{-s})^{-1}.$$

This is well-defined and gives the Euler factors at all finite primes.

Remark. Since G is a finite group, once ρ is given, there are only a finite number of characteristic polynomials that can occur. For example, if we take the trivial one-dimensional representation, only the polynomial $(1 - T)$ occurs. But the subtlety in the Artin L-function is the assignment $\mathfrak{p} \mapsto \sigma_{\mathfrak{p}}$. In other words, which one of the finite number of characteristic polynomials is assigned to a given prime $\mathfrak{p}$ determines and is determined by the arithmetic of the field extension, in particular the splitting of primes.

We have also to define the Archimedean Euler factors. For each Archimedean prime v of K we set

$$L_v(s,\chi,K) = \begin{cases} ((2\pi)^{-s}\Gamma(s))^{\chi(1)} & \text{if } v \text{ is complex} \\ ((\pi^{-s/2}\Gamma(s/2))^a(\pi^{-(s+1)/2}\Gamma((s+1)/2))^b & \text{if } v \text{ is real.} \end{cases}$$

Here

$$a + b = \chi(1)$$

and a (resp. b) is the dimension of the $+1$ eigenspace (resp. -1 eigenspace) of complex conjugation.

We shall write

$$\gamma(s,\chi,K) = \prod_{v \text{ infinite}} L_v(s,\chi,K).$$

The Artin L-function $L(s,\chi,K)$ satisfies a functional equation of the following type. First, one defines the Artin conductor $\mathfrak{f}_\chi$ associated to χ. It is an ideal of K and is defined in terms of the restriction of χ to the inertia groups and its various subgroups.

More precisely, let ν be a place of K. Let w be a place of L dividing ν and let G_0 denote the inertia group I_w at w. We have a descending filtration of higher ramification groups (see [CF], p. 33]).

$$G_0 \supseteq G_1 \supseteq \cdots.$$

Let V be the underlying representation space for ρ. Define

$$n(\chi,\nu) = \sum_{i=0}^{\infty} \frac{|G_i|}{|G_0|} \operatorname{codim}(V^{G_i}).$$

Then $n(\chi,\nu)$ is an integer and is well-defined (that is, it is independent of the choice of w above ν). Moreover, it is equal to zero apart from a finite number of ν. This allows us to define the ideal

$$\mathfrak{f}_\chi = \prod_\nu \mathfrak{p}_\nu^{n(\chi,\nu)}.$$

We also set

$$A_\chi = d_K^{\chi(1)} \mathbf{N}_{K/\mathbb{Q}} \mathfrak{f}_\chi.$$

Let us set

$$\Lambda(s,\chi,K) = A_\chi^{s/2}\gamma(s,\chi,K)L(s,\chi,K).$$

Then we have the functional equation

$$\Lambda(s,\chi,K) = W(\chi)\Lambda(1-s,\bar{\chi},K)$$

where $W(\chi)$ is a complex number of absolute value 1.

The number $W(\chi)$ itself carries deep arithmetic information. For example, it is related to Galois module structure. The reader is referred to the monograph [Fr] of Fröhlich for an introduction to this subject.

We now recall some of the formalism of Artin L-functions and their basic properties. It is summarized in the two properties:

$$L(s, \sum_{\chi} a_\chi \chi, K) = \prod_{\chi} L(s, \chi, K)^{a_\chi} \qquad \text{for any} \qquad a_\chi \in \mathbb{Z} \tag{1}$$

$$L(s, \operatorname{Ind}_H^G \chi, K) = L(s, \chi, L^H) \text{ where } L^H \text{ is the subfield of } L \text{ fixed by } H. \tag{2}$$

Using (1) and (2), we find that

$$\prod_{\chi \text{ irred}} L(s, \chi, K)^{\chi(1)} = L(s, \operatorname{reg}_G, K) = L(s, 1, L) = \zeta_L(s)$$

$$= \prod_{\mathfrak{q}} (1 - (\mathbf{N}\mathfrak{q})^{-s})^{-1}.$$

There is a theorem of Brauer which says that for any irreducible χ, there are subgroups $\{H_i\}$, one-dimensional characters ψ_i of H_i and integers $m_i \epsilon \, \mathbb{Z}$ with

$$\chi = \sum_i m_i \operatorname{Ind}_{H_i}^G \psi_i.$$

Using (1) and (2), we see that

$$L(s, \chi, K) = \prod_i L(s, \psi_i, L^{H_i})^{m_i}.$$

If χ is one-dimensional, then Artin's reciprocity theorem identifies $L(s, \chi, K)$ with a Hecke L-series for a ray class character. By Hecke and Tate, we know the analytic continuation of these L-series (see Chapters 13 and 14 of [La]).

From the Brauer induction theorem, it follows that any Artin L-function has a meromorphic continuation. **Artin's conjecture** asserts that every Artin L function $L(s, \chi, K)$ associated to a character χ of $\operatorname{Gal}(\bar{K}/K)$ has an analytic continuation for all s except possibly for a pole at $s = 1$ of order equal to the multiplicity of the trivial representation in ρ. (Note that χ determines ρ up to isomorphism and so our notation is justified).

This is a very central and important conjecture in number theory. It is part of a general reciprocity law. The conjecture of Artin is known to hold in many cases. Most of these arise from a combination of the one-dimensional case and group theory. Some examples are given in the exercises.

Returning to the general case, we see from the factorization

$$\zeta_L(s) = \prod_{\chi \text{ irred}} L(s, \chi, K)^{\chi(1)}$$

that Artin's conjecture implies that $\zeta_L(s)/\zeta_K(s)$ is entire. In fact, let L/K be a (not necessarily Galois) finite extension and let $\tilde{K}/K$ be its normal closure. Say $G = \mathrm{Gal}(\tilde{K}/K)$ and $H = \mathrm{Gal}(\tilde{K}/L)$. Then

$$L(s, \mathrm{Ind}_H^G(1_H), K) = L(s, 1_H, L) = \zeta_L(s).$$

On the other hand,

$$\mathrm{Ind}_H^G 1_H = 1_G + \sum_{1 \neq \chi \text{ irred}} a_\chi \chi$$

with $0 \leq a_\chi \in \mathbb{Z}$. So,

$$L(s, \mathrm{Ind}_H^G 1_H, K) = \zeta_K(s) \prod L(s, \chi, K)^{a_\chi}.$$

Putting these together, we see that Artin's conjecture implies that $\zeta_L(s)/\zeta_K(s)$ is entire, whether L/K is Galois or not. This special case of Artin's conjecture is called Dedekind's conjecture. Below, we shall discuss it in several cases. In particular, it is known to hold in the case L/K is Galois (Aramata-Brauer) and in case $\tilde{L}/K$ is solvable (Uchida-van der Waall).

§3 The Aramata-Brauer Theorem

Let L/K be Galois with group G.

Theorem 3.1 *The quotient $\zeta_L(s)/\zeta_K(s)$ is entire.*

By the properties of Artin L-functions described in §2, the Theorem follows from the following result.

Proposition 3.2 *There are subgroups $\{H_i\}$, 1-dimensional character ψ_i of H_i and $0 \leq m_i \in \mathbb{Z}$ so that*
$$\mathrm{reg}_G - 1_G = \sum m_i \mathrm{Ind}_{H_i}^G \psi_i.$$

(Note that $(\mathrm{reg}_G, 1_G) = (\mathrm{Ind}_{\{e\}}^G 1, 1_G) = (1, 1_G|_{\{e\}}) = 1$ by Frobenius reciprocity).

For any cyclic subgroup A define $\theta_A : A \to \mathbb{C}$ by

$$\theta_A(\sigma) = \begin{cases} |A| & \text{if } \sigma \text{ generates } A \\ 0 & \text{else} \end{cases}$$

and

$$\lambda_A = \phi(|A|) \mathrm{reg}_A - \theta_A,$$

where ϕ denotes Euler's function.

Thus,

$$\lambda_A(\sigma) = \begin{cases} \phi(|A|)|A| & \text{if } \sigma = 1 \\ -\theta_A(\sigma) & \text{if } \sigma \neq 1 \end{cases}$$

Proposition 3.2 will be proved in two steps.

Step 1. $\lambda_A = \sum m_\chi \chi$ with $m_\chi \geq 0$, $m_\chi \in \mathbb{Z}$ and χ ranges over the characters of A.

Step 2. $\mathrm{reg}_G - 1_G = \frac{1}{|G|} \sum_A \mathrm{Ind}_A^G \lambda_A$ where the sum is over all cyclic subgroups A of G.

To prove Step 1, it is enough to show that $(\lambda_A, \chi) \geq 0$ for any irreducible χ of A. But

$$(\lambda_A, \chi) = \phi(|A|) - (\theta_A, \chi)$$
$$= \phi(|A|) - \sum_{\substack{\sigma \in A \\ <\sigma>=A}} \chi(\sigma) = \sum_{\substack{\sigma \in A \\ <\sigma>=A}} (1 - \chi(\sigma))$$
$$= Tr(1 - \chi(\sigma)) \in \mathbb{Z} \text{ (for any generator } \sigma \text{ of } A)$$

Now for $\chi \neq 1$, $\mathrm{Re}(1 - \chi(\sigma)) > 0$ if $\sigma \neq e$ and $= 0$ if $\sigma = e$. Then, if $A \neq \{1\}$, (λ_A, χ) is positive for all $\chi \neq 1$ and $= 0$ if $\chi = 1$. If $A = \{1\}$ then $\lambda_A = 0$. This proves Step 1.

To prove the equality of Step 2, it is enough to show that for any irreducible character ψ of G, both sides have the same inner product with ψ. Now

$$(|G|(\mathrm{reg}_G - 1_G), \psi) = \sum (\mathrm{reg}_G - 1_G)(g)\overline{\psi(g)}$$
$$= |G|\psi(1) - \sum_{g \in G} \psi(g)$$

Also, by Frobenius reciprocity,

$$\sum_A (\mathrm{Ind}_A^G \lambda_A, \psi) = \sum_A (\lambda_A, \psi|_A)$$
$$= \sum_A \{\phi(|A|)\psi(1) - \sum_{\substack{\sigma \in A \\ <\sigma>=A}} \psi(\sigma)\}$$
$$= \psi(1) \sum_A \phi(|A|) - \sum_{\sigma \in G} \psi(\sigma).$$

Now

$$\sum_A \phi(|A|) = \sum_A \sum_{\substack{\sigma \in A \\ <\sigma>=A}} 1 = \sum_{\sigma \in G} 1 = |G|.$$

This completes Step 2 and the proof of Proposition 3.2.

We illustrate the equality of Step 2 above with an example. Let $L/\mathbb{Q}$ be a biquadratic extension (Galois). Then the identity is

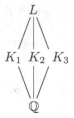

$$\left(\frac{\zeta_L(s)}{\zeta(s)}\right)^4 = \left(\frac{\zeta_L(s)}{\zeta_{K_1}(s)}\right)^2 \left(\frac{\zeta_L(s)}{\zeta_{K_2}(s)}\right)^2 \left(\frac{\zeta_L(s)}{\zeta_{K_3}(s)}\right)^2$$

which when unwound, gives the usual factorization

$$\zeta_L(s) = \zeta(s)L(s,\chi_1)L(s,\chi_2)L(s,\chi_3).$$

§4 Dedekind's conjecture in the non-Galois case

As explained in §2, Artin's holomorphy conjecture implies that the quotient $\zeta_K(s)/\zeta_F(s)$ is entire even when K/F is *not* normal. This latter assertion, called Dedekind's conjecture, is still an open problem in general.

Dedekind's conjecture has been settled in a few cases, notably for extensions K/F whose normal closure has solvable Galois group. This is due to Uchida [Uc] and van der Waall [vdW]. In fact, their method allows us to prove the following.

Theorem 4.1 *Let K/F be a finite extension of number fields and suppose that the normal closure $\tilde{K}/F$ has Galois group G which is the semidirect product of $H = \mathrm{Gal}(\tilde{K}/K)$ by an abelian normal subgroup A of G. Then Dedekind's conjecture is true for K/F. That is,*

$$\zeta_K(s)/\zeta_F(s)$$

is entire.

Proof. Let us write

$$\mathrm{Ind}_H^G(1_H) = \sum m_\chi \chi$$

where $0 \leq m_\chi \in \mathbb{Z}$, $m_1 = 1$ and χ ranges over the irreducible characters of G. Consider

$$\mathrm{Ind}_H^G(1_H)|_A = \sum m_\chi \chi|_A.$$

By Mackey's theorem,

$$\mathrm{Ind}_H^G(1_H)|_A = \mathrm{Ind}_{H \cap A}^A(1_H|_{H \cap A}) = \mathrm{Ind}_{\{1\}}^A 1 = \mathrm{reg}_A.$$

Thus,

$$\operatorname{Ind}_H^G(1_H)|_A = \sum \epsilon$$

where ϵ ranges over all the irreducible characters of A. Thus, for all χ,

$$m_\chi = 0 \text{ or } 1 \text{ and } (\chi|_A, \epsilon) = 0 \text{ or } 1 \text{ for any } \epsilon \in Irr(A).$$

Now, take an $\epsilon \in Irr(A)$ such that there is a $\chi \in Irr(G)$ with $m_\chi \neq 0$ and $(\chi|_A, \epsilon) = 1$. Let T_ϵ be the inertia group of ϵ :

$$T_\epsilon = \{\sigma \in G : \epsilon^\sigma = \epsilon\}.$$

(Here, ϵ^σ is the character $a \mapsto \epsilon(\sigma a \sigma^{-1})$.) Of course, $T_\epsilon \supseteq A$ and we can write it as $T_\epsilon = H_\epsilon A$ where $H_\epsilon \subseteq H$. We can extend ϵ to a character $\tilde\epsilon$ of T_ϵ by setting

$$\tilde\epsilon(ha) = \epsilon(a) \text{ for any } h \in H_\epsilon, a \in A.$$

Let us write

$$\operatorname{Ind}_A^{T_\epsilon} \epsilon = \operatorname{Ind}_A^{T_\epsilon} \tilde\epsilon|_A = \sum_{\psi \in Irr(T_\epsilon)} m_\psi \psi.$$

Notice that

$$\operatorname{Ind}_A^{T_\epsilon} \epsilon(g) = \begin{cases} [T_\epsilon : A]\epsilon(g) & \text{if } g \in A \\ 0 & \text{otherwise} \end{cases}.$$

Thus,

$$[T_\epsilon : A]\epsilon = (\operatorname{Ind}_A^{T_\epsilon} \epsilon)|_A = \sum m_\psi \psi|_A.$$

Thus for every $\psi \in Irr(T_\epsilon)$, with $m_\psi \neq 0$, $\psi|_A$ is a multiple of ϵ. In fact,

$$m_\psi = (\psi, \operatorname{Ind}_A^{T_\epsilon} \epsilon) = (\psi|_A, \epsilon),$$

and so $\psi|_A = m_\psi \epsilon$. It follows that

$$\sum m_\psi^2 = [T_\epsilon : A].$$

From this, we deduce that the characters $\{\operatorname{Ind}_{T_\epsilon}^G \psi\}$ are distinct and irreducible. Indeed, we have

$$[T_\epsilon : A] = \sum m_\psi^2 \leq (\operatorname{Ind}_A^G \epsilon, \operatorname{Ind}_A^G \epsilon) = (\epsilon, (\operatorname{Ind}_A^G \epsilon)|_A)$$

and

$$\operatorname{Ind}_A^G \epsilon|_A = [T_\epsilon : A] \sum \epsilon^g$$

where the sum on the right is over a set of coset representatives for T_ϵ in G. By definition of T_ϵ, the conjugates ϵ^g are distinct and our claim follows. Now,

$$1 = (\chi|_A, \epsilon) = (\chi, \operatorname{Ind}_A^G \epsilon) = \sum_{\psi \in Irr(T_\epsilon)} m_\psi(\chi, \operatorname{Ind}_{T_\epsilon}^G \psi).$$

Thus, there is a unique $\phi = \phi(\chi) \in Irr(T_\epsilon)$ with $m_\phi = 1$ and $(\chi, \operatorname{Ind}_{T_\epsilon}^G \phi) = 1$. By the irreducibility of both characters, it follows that $\chi = \operatorname{Ind}_{T_\epsilon}^G \phi$. Also, as $\psi|_A = m_\psi \epsilon$, we have $\phi(1) = m_\phi = 1$. Hence, χ is the induction of a linear character. This proves that $\operatorname{Ind}_H^G 1_H$ is a sum of monomial characters and the proposition follows.

Corollary 4.2   (Uchida, van der Waall) *Let K/F be an extension of number fields and $\tilde{K}$ a normal closure of K/F. Suppose that $\mathrm{Gal}(\tilde{K}/F)$ is solvable. Then $\zeta_K(s)/\zeta_F(s)$ is entire.*

Proof. As above, we set $G = \mathrm{Gal}(\tilde{K}/F)$ and $H = \mathrm{Gal}(\tilde{K}/K)$, We proceed by induction on the order of $|G|$. We may assume that H is a maximal subgroup of G. For if J is a maximal subgroup of G with $H \subset J \subset G$, and M is the fixed field of J, then

$$\zeta_K(s)/\zeta_F(s) = (\zeta_K(s)/\zeta_M(s))\,(\zeta_M(s)/\zeta_F(s))$$

where the first factor on the right is entire by the induction hypothesis and the second by the maximality of J.

Also, since G corresponds to the normal closure of K/F, we may assume that H does not contain any proper non-trivial normal subgroup of G. Now let A be a minimal normal subgroup of G. As G is solvable, such an A exists and is (elementary) abelian. Moreover, A is not contained in H. Then $HA = G$ and $H \cap A = \{1\}$. Indeed, the first equality is just the maximality of H and the second follows from the minimality of A and the observation that $H \cap A$ is again a normal subgroup. Thus, G is the semidirect product of A by H and Theorem 4.1 applies.

Finally in this section, we can ask the following variant of Dedekind's conjecture. Let L/K be an extension with group G, and let H be a subgroup. Let ρ be an irreducible representation of G. Then, is the quotient

$$L(s, \mathrm{Ind}_H^G(\rho|_H), K)/L(s, \rho, K) \tag{$\ddagger$}$$

entire? This includes the general case of Dedekind's conjecture (if we take $\rho = 1_G$). ($\ddagger$) can be proved by the method of the Proposition above, if G is solvable and ρ is an abelian character. Indeed, we need only make two observations. First, if we write

$$\mathrm{Ind}_H^G(\rho|_H) - \rho = \sum m_\chi \chi$$

then restricting to A shows that

$$\sum m_\chi \chi|_A = \rho(1) \sum \epsilon.$$

Moreover, if G is any group and A is an abelian normal subgroup, and ϵ is an (irreducible) character of A, then

$$\mathrm{Ind}_A^G \epsilon = \sum m_i \, \mathrm{Ind}_{T_\epsilon}^G \psi_i$$

where $\psi_i(1) = m_i$ and T_ϵ is the inertia subgroup of ϵ. Thus, if $\rho(1) = 1$, and $(\chi|_A, \epsilon) \neq 0$, then

$$1 = (\chi|_A, \epsilon) = (\chi, \mathrm{Ind}_A^G \epsilon) = \sum m_i(\chi, \mathrm{Ind}_{T_\epsilon}^G \psi_i)$$

and as before, this implies that there is an i with $m_i = 1$ and $\chi = \operatorname{Ind}_{T_\epsilon}^G \psi_i$. Even without assuming $\rho(1) = 1$, we get $\chi = \operatorname{Ind}_{T_\epsilon}^G \psi_i$ for some ϵ and i. But we may not know the holomorphy of $L(s, \psi_i)$. (Notice also, that we can restrict to the case H is a maximal subgroup. For if $J \supseteq H$ is a maximal subgroup, ($M =$ fixed field of J)

$$L(s, \operatorname{Ind}_H^G(\rho|_H), K)/L(s, \operatorname{Ind}_J^G(\rho|_J), K) = L(s, \operatorname{Ind}_H^J(\rho|_H), M)/L(s, \rho|_J, M).$$

and so

$$\frac{L(s, \operatorname{Ind}_H^G(\rho|_H), K)}{L(s, \rho, K)} = \frac{L(s, \operatorname{Ind}_H^J(\rho|_H), M)}{L(s, \rho|_J, M)} \cdot \frac{L(s, \operatorname{Ind}_J^G(\rho|_J), K)}{L(s, \rho, K)}.$$

§5 Zeros and poles of Artin L-functions

There is another approach to the Aramata-Brauer theorem which does not explicitly use the decomposition of $\operatorname{reg}_G - \chi$ into monomial characters. To describe it, let us set

$$n_\chi = n_\chi(s_0) = ord_{s=s_0} L(s, \chi, F).$$

Then, in [St], the inequality

$$\sum_{\chi \in Irr(G)} n_\chi^2 \le r^2, \quad r = ord_{s=s_0} \zeta_K(s)$$

is proved. From this, it follows for example that $\zeta_K(s)/L(s, \chi, F)$ is entire except possibly at $s = 1$, and that the same holds for the product $\zeta_K(s)L(s, \chi, F)$. This raises the question of whether $\operatorname{reg}_G - \chi$ can be decomposed as a non-negative sum of monomial characters. This was answered in the affirmative by Rhoades [R]. Some special cases were computed in [Mu1].

Our approach applies in a wider context of an L-function formalism which is satisfied by a variety of objects in number theory and algebraic geometry. Let G be a finite group. For every subgroup H of G and complex character ψ of H, we attach a complex number $n(H, \psi)$ satisfying the following properties:

(1) Additivity: $n(H, \psi + \psi') = n(H, \psi) + n(H, \psi')$,
(2) Invariance under induction: $n(G, \operatorname{Ind}_H^G \psi) = n(H, \psi)$.

The formalism can be applied to the above case when G is the Galois group of a normal extension K/k and $n(H, \psi)$ is the order of the zero at $s = s_0$ of the Artin L-series attached to ψ corresponding to the Galois extension K/K^H. It can also be applied to the situation when E is an elliptic curve over k and $n(H, \psi)$ corresponds to the order of the zero at $s = s_0$ of the "twist" by ψ of the L-function of E over K^H (see [MM] for definitions and details).

We consider the following generalized character introduced by Heilbronn:

$$\theta_H = \sum_\psi n(H,\psi)\psi$$

where the sum is over all irreducible characters ψ of H. Our first step is to show that

Proposition 5.1 $\theta_G|_H = \theta_H$.

Proof.

$$\theta_G|_H = \sum_\chi n(G,\chi)\chi|_H$$

$$= \sum_\chi n(G,\chi)\left(\sum_\psi (\chi|_H,\psi)\psi\right)$$

where the inner sum is over all irreducible characters of H and the outer sum is over all irreducible characters of G. By Frobenius reciprocity, $(\chi|_H,\psi) = (\chi,\mathrm{Ind}_H^G\psi)$ and so

$$\theta_G|_H = \sum_\psi \left(\sum_\chi n(G,\chi)(\chi,\mathrm{Ind}_H^G\psi)\right)\psi.$$

But now, by property (1), the inner sum is $n(G,\mathrm{Ind}_H^G\psi)$ which equals $n(H,\psi)$ by property (2). Thus, $\theta_G|_H = \theta_H$.

This immediately implies:

Proposition 5.2   *Let* reg *denote the regular representation of G. Suppose for every cyclic subgroup H of G, we have $n(H,\psi) \geq 0$. Then $n(G,\chi)$ is real for every irreducible character χ of G and*

$$\sum_\chi n(G,\chi)^2 \leq n(G,\mathrm{reg})^2.$$

Proof. By Artin's theorem, every character can be written as a rational linear combination of characters induced from cyclic subgroups and so $n(G,\chi)$ is real. By the orthogonality relations,

$$(\theta_G,\theta_G) = \sum_\chi n(G,\chi)^2.$$

On the other hand,

$$(\theta_G,\theta_G) = \frac{1}{|G|}\sum_{g\in G} |\theta_G(g)|^2.$$

By Proposition 5.1,

$$\theta_G(g) = \theta_{\langle g \rangle}(g) = \sum_\psi n(\langle g \rangle, \psi) \psi(g)$$

which is bounded by $n(G, \text{reg})$ in absolute value by our hypothesis and property (1). This completes the proof.

Similar reasoning implies

Proposition 5.3 *Let ρ be an arbitrary character of G. Suppose for every cyclic subgroup H of G, and irreducible character ψ of H, we have $n(H, \rho|_H \otimes \psi) \geq 0$, then $n(G, \rho \otimes \chi)$ is real for every irreducible character χ of G and*

$$\sum_\chi n(G, \rho \otimes \chi)^2 \leq n(G, \rho \otimes \text{reg})^2.$$

These results can also be generalized to the context of automorphic forms. Some preliminary work in this direction can be found in [MM].

§6 Low order zeros of Dedekind zeta functions

By analogy with the conjecture that the zeros of the Riemann zeta function are simple, one expects that the n_χ are bounded. One might ask whether

$$n_\chi \ll \chi(1)$$

or even the stronger

$$n_\chi \ll 1$$

holds.

We begin by establishing a zero-free region for Dedekind zeta functions. This is due to Stark [St]. This in turn gives a region where Artin L-functions are zero-free except possibly for a simple exceptional zero.

Proposition 6.1 *Let M be an algebraic number field of degree $n = r_1 + 2r_2$ where M has r_1 real embeddings and $2r_2$ complex conjugate embeddings. For $\sigma > 1$ we have*

$$-\frac{\zeta_M'}{\zeta_M}(\sigma) < \frac{1}{\sigma} + \frac{1}{\sigma - 1} + \frac{1}{2}\log\left(\frac{|d_m|}{2^{2r_2}\pi^n}\right) + \frac{r_1}{2}\frac{\Gamma'}{\Gamma}(\sigma/2) + r_2\frac{\Gamma'}{\Gamma}(\sigma).$$

Also, if $M \neq \mathbb{Q}$, ζ_M has at most one zero in the region

$$\sigma \geq 1 - \frac{1}{4\log|d_m|}, \quad |t| \leq \frac{1}{4\log|d_m|}.$$

Proof. Consider $f(s) = s(s-1)\zeta_M(s)$. By logarithmically differentiating the Hadamard factorization, we get the relation

$$\sum_\rho \frac{1}{s-\rho} = \frac{1}{s-1} + \frac{1}{2}\log|d_m|$$

$$+ \left(\frac{1}{s} - \frac{n}{2}\log\pi\right) + \frac{r_1}{2}\frac{\Gamma'}{\Gamma}(\tfrac{s}{2}) + r_2\left(\frac{\Gamma'}{\Gamma}(s) - \log 2\right) + \frac{\zeta'_M}{\zeta_M}(s).$$

The sum on the left runs over zeros ρ of $\zeta_M(s)$ in the strip $o < \sigma < 1$ and the terms with ρ and $\bar\rho$ are grouped together.

For $s = \sigma > 1$ we have

$$\frac{1}{\sigma-\rho} + \frac{1}{\sigma-\bar\rho} > 0.$$

Thus for $\sigma > 1$ we have

$$\sum_\rho{}' \frac{1}{\sigma-\rho} \leq \sum_\rho \frac{1}{\sigma-\rho}$$

where the sum on the left denotes summation over any convenient subset of the zeros ρ which is closed under complex conjugation. In particular, the sum

$$\sum \frac{1}{\sigma-\rho}$$

is positive and we deduce the inequality of the Proposition.

Now take $s = \sigma$ with $1 < \sigma < 2$. All the terms on the right of the above inequality after $\frac{1}{2}\log|d_m|$ are negative and thus

$$\sum_\rho{}' \frac{1}{\sigma-\rho} < \frac{1}{\sigma-1} + \frac{1}{2}\log|d_m|.$$

If $\rho = \beta + i\gamma$ is in the rectangle specified in the statement, (with $\gamma \neq 0$) then $\bar\rho$ is also in the same rectangle and taking the contribution from ρ and $\bar\rho$ only, we get the inequality

$$\frac{2(\sigma-\beta)}{(\sigma-\beta)^2 + \gamma^2} < \frac{1}{\sigma-1} + \frac{1}{2}\log|d_m|.$$

But this is false for $M \neq \mathbb{Q}$ at $\sigma = 1 + \frac{1}{\log|d_m|} < 2$. The same value of σ gives a contradiction if there are two real zeroes in this rectangle (or a single real multiple zero). This completes the proof of the Proposition.

The following consequence is also due to Stark [St].

Corollary 6.2 *Let K/F be a Galois extension. For any Artin L-function $L(s, \chi, F)$ of this extension, the region*

$$\sigma \geq 1 - \frac{1}{4 \log |d_k|}, \quad |t| \leq \frac{1}{4 \log |d_k|}.$$

is free of zeros except possibly for a simple zero. This zero exists only if χ is a real Abelian character of a quadratic subfield of K.

Next, we examine the case when the Dedekind zeta function may in fact vanish, but the order of zero is small. We shall study this under the assumption that K/F itself is a solvable Galois extension. If we are at a point $s = s_0$ where $\zeta_K(s)$ has a "small" order zero, then it is possible to show more than just the analyticity of $\zeta_K(s)/\zeta_F(s)$ at s_0. We have the following result due to Foote and K. Murty [FM].

Theorem 6.3 *Let K/F be a solvable extension and write*

$$[K : F] = p_1^{\alpha_1} \cdots p_t^{\alpha_t}, \quad p_1 < p_2 < \cdots < p_t$$

for the prime power decomposition of the degree. Suppose that at $s = s_0$, we have

$$r = ord_{s=s_0} \zeta_K(s) \leq p_2 - 2.$$

Then for each $\chi \in Irr(G)$, the Artin L-series $L(s, \chi, F)$ is analytic at $s = s_0$.

This has the following immediate corollary.

Corollary 6.4 *If K/F is a Galois extension of odd degree and $\zeta_K(s)$ has a zero of order ≤ 3 at a point s_0 then all Artin L-functions of K/F are analytic at s_0.*

This represents a partial generalization of the result Corollary 6.2 of Stark. Of course, Stark's result makes no assumption on the Galois group of K/F.

We give a brief outline of the proof. Assume the theorem is false, and take G to be a minimal counterexample for which Artin's conjecture fails, at a point $s = s_0$ where the order of $\zeta_K(s)$ is small as explained in the statement. We want to prove that the generalized character θ_G defined above is an actual character. We repeatedly use the two key properties of θ_G namely,

$$\theta_G|_H = \theta_H \text{ for any subgroup } H \text{ of } G$$

and

$$\theta_G(1) = ord_{s=s_0} \zeta_K(s).$$

The first follows from Proposition 5.1 and the second follows from the factorization of ζ_K into the $L(s, \chi, F)$. Moreover, by our assumption of minimality, we may suppose that θ_H is a character for every proper subgroup H of G. In addition, the induction hypothesis and the invariance of L-functions under induction allow us to assume that χ is not induced from any proper subgroup of G. Also, we may assume that χ is faithful. For if $\operatorname{Ker}\chi$ is non-trivial and M (say) denotes its fixed field, then by the Aramata-Brauer theorem (Theorem 3.1), $\zeta_M(s)$ divides $\zeta_K(s)$. In particular, $ord_{s=s_0}\zeta_M(s) \le r$ and the second smallest prime divisor of $[M : F]$ is $\ge p_2$. Since $L(s,\chi,F)$ is the same whether viewed as an L-function of K or M, the analyticity of this L-function at s_0 would follow from the induction hypothesis.

We now decompose θ_G into three parts $\theta_1, \theta_2, \theta_3$ as follows. Let θ_3 be the sum of all terms $n_\lambda \lambda$ such that λ is *not* a faithful character of G. Let $-\theta_2$ be the sum of all terms $n_\chi \chi$ for which n_χ is *negative*. Finally, let θ_1 be the sum of all terms $n_\psi \psi$ where ψ is a faithful character with $n_\psi > 0$. Again by the assumption of minimality, we see that $(\theta_2, \theta_3) = 0$ and by definition, θ_1 is orthogonal to θ_2 and θ_3. Thus, we get the decomposition

$$\theta_G = \theta_1 - \theta_2 + \theta_3.$$

We will now get further information about the constituents of θ_2 by restricting to an appropriate subgroup. As we shall see, a key tool in this is Clifford's theorem. It provides us with two pieces of information.

Firstly, since G is solvable and non-abelian, it has a normal subgroup N of prime index, p say, which contains the center $Z(G)$ of G. Clifford's theorem tells us that for any $\chi \in Irr(G)$, $\chi|_N$ is either irreducible or χ is the induction of a character from N. In particular, if we take for χ a summand of θ_2, it follows that $\chi|_N$ is irreducible.

Secondly, it tells us that any abelian normal subgroup must be central (that is, contained in the center), for otherwise every $\chi \in Irr(G)$ would be induced from a proper subgroup contradicting the non-triviality of θ_2.

Now, *every* non-trivial normal subgroup of a solvable group contains a non-trivial abelian subgroup which is normal in G. Thus, no irreducible constituent λ of θ_3 is faithful on the center. We must therefore have

$$(\theta_2|_N, \theta_3|_N) = 0.$$

Since

$$\theta_G|_N = \theta_1|_N - \theta_2|_N + \theta_3|_N$$

is a character of N, it follows that

$$\text{either } \theta_1|_N = \theta_2|_N \text{ or } \theta_1|_N = \theta_2|_N + \phi$$

for some character ϕ of N. A further argument using Clifford's theorem in fact eliminates the second possibility. Indeed, choose an irreducible component α of $\phi = \theta_1|_N - \theta_2|_N$ and let ψ be an irreducible component of $\theta_1 - \theta_2$ such that $\psi|_N$ contains α. Notice that the G conjugates of α are also contained in $\theta_1|_N - \theta_2|_N$

and hence also in $\psi|_N$. It follows that the sum of the distinct conjugates form a character of degree $\leq \phi(1)$. Clifford's theorem tells us that $\psi|_N$ is equal to this sum and so $\psi(1) \leq \phi(1) \leq r$. Also, ψ is a constituent of θ_1 and so it is faithful by definition.

Now, a theorem of Ito tells us that in a solvable group, a p-Sylow subgroup is abelian and normal if there is a faithful character of degree smaller than $p - 1$. Thus, the conclusion of the previous paragraph and our assumption that $r \leq p_2 - 2$ imply that G has an abelian normal subgroup of order $n/p_1^{\alpha_1}$. This would force G to be nilpotent and Artin's conjecture is known to hold for such groups as every irreducible character is monomial. This again contradicts the nontriviality of θ_2. We conclude that $\theta_1|_N = \theta_2|_N$. This is the only step in which the assumed bound on r is used.

The final contradiction now comes by showing that $\theta_1 = \theta_2$. To do this, take $x \in G \backslash N$. Denote by H the subgroup generated by x and the center of G. As H is abelian, it is a proper subgroup of G. As we observed earlier, every irreducible component λ of θ_3 has the property that its kernel $\operatorname{Ker} \lambda$ meets the center non-trivially. Thus, the same holds for $\operatorname{Ind}_H^G(\lambda|_H)$. Now, taking an irreducible component χ of θ_2, we know that χ is faithful and so

$$(\chi, \operatorname{Ind}_H^G(\lambda|_H)) = 0.$$

By Frobenius reciprocity, $(\chi|_H, \lambda|_H) = 0$ and so

$$(\theta_2|_H, \theta_3|_H) = 0.$$

Now,

$$\theta_G|_H = \theta_1|_H - \theta_2|_H + \theta_3|_H$$

and again $\theta_G|_H$ is a character of H. Thus, $\theta_1|_H - \theta_2|_H$ is either zero or a character. By our earlier argument, we know that $\theta_1(1) = \theta_2(1)$ and so we must have $\theta_1|_H = \theta_2|_H$. Combined with our earlier result for N it follows that

$$\theta_1 = \theta_2.$$

This contradiction completes the proof.

The argument suggests that the condition $r \leq p_2 - 2$ be replaced by a bound on r involving the least degree of a faithful character. Results in this direction have in fact now been obtained by Foote [Fo] and by Foote and Wales [FW].

§7 Chebotarev density theorem

Let K/F be a finite Galois extension of number fields with group G. Let C be a subset of G which is stable under conjugation. Thus C is a union of conjugacy classes. Define

$$\pi_C(x) = \#\{\nu \text{ a place of } F \text{ unramified in } K, \ \mathbf{N}_{F/\mathbb{Q}}(\mathfrak{p}_\nu) \leq x \text{ and } \sigma_\nu \subset C\}.$$

The Chebotarev density theorem asserts that

$$\pi_C(x) \sim \frac{|C|}{|G|}\pi_F(x)$$

where $\pi_F(x)$ denotes the number of primes of F of norm $\leq x$. Effective versions of this theorem were given by Lagarias and Odlyzko [LO]. We state two of their results. The first of these assumes the Riemann Hypothesis for Dedekind zeta functions. The second is unconditional.

Theorem 7.1 *Suppose the Dedekind zeta function $\zeta_K(s)$ satisfies the Riemann hypothesis. Then*

$$\pi_C(x) = \frac{|C|}{|G|}\pi_F(x) + \mathbf{O}(\frac{|C|}{|G|} \cdot x^{\frac{1}{2}}(\log d_L + n_L \log x)).$$

This version of their result is only slightly more refined than the statement given in [LO] and is due to Serre [Se2, p. 133]. The proof of Theorem 7.1 is very analogous to the classical proof of the prime number theorem in arithmetic progressions, as presented, for example, in the monograph of Davenport [D]. However, there are some points of difference and we now briefly discuss them.

As in the classical case, the proof begins by expressing the characteristic function of the conjugacy class C in terms of characters of G. However, we have to deal with the fact that G is non-abelian and that we do not know the analytic properties of Artin L-functions. In particular, we do not know Artin's conjecture. We have

$$\delta_C = \frac{|C|}{|G|}\sum_\chi \overline{\chi(g_C)}\chi$$

where δ_C denotes the characteristic function of the class C and g_C is any element in this class. Hence

$$\pi(x,\delta_C) = \frac{|C|}{|G|}\sum_\chi \overline{\chi(g_C)}\pi(x,\chi)$$

where for any class function ϕ we set

$$\pi(x,\phi) = \sum_{N\nu \leq x} \phi(\sigma_\nu).$$

Here the sum is over places ν of F unramified in K and of norm $\leq x$.

If we want to include ramified primes and also prime powers in the sums, we introduce the function

$$\tilde{\pi}(x,\phi) = \sum_{N\nu^m \leq x} \phi(\sigma_\nu^m)$$

where in the case ν is a ramified prime, we define

$$\phi(\sigma_\nu^m) = \frac{1}{|I_w|} \sum \phi(g)$$

where I_w is the inertia group at a prime w of K dividing ν and the sum is over elements g in the decomposition group D_w whose image in the quotient D_w/I_w maps to σ_ν^m. The advantage in this sum $\tilde{\pi}$ is that it is closely related to the logarithmic derivative of the Artin L function.

At this point, we use some group theory to replace the Artin L-functions with Hecke L-functions. Indeed, let H be a subgroup of G and h an element of H. Let C_H denote its conjugacy class in H and C its conjugacy class in G. Let

$$\delta : H \longrightarrow \{0, 1\}$$

denote the characteristic function of C_H. Now set

$$\phi = \operatorname{Ind}_H^G \delta.$$

By definition, we see that ϕ is supported only on the conjugacy class C and so $\phi = \lambda \delta_C$. The value of λ is easily computed by Frobenius reciprocity:

$$\lambda \frac{|C|}{|G|} = (\phi, 1_G) = (\delta, 1_H) = \frac{|C_H|}{|H|}.$$

Thus

$$\lambda = \frac{|C_H| \cdot |G|}{|H| \cdot |C|}.$$

From the inductive property of L-functions, it is not hard to see that

$$\tilde{\pi}(x, \phi) = \tilde{\pi}(x, \delta).$$

Now the right hand side is written as a sum involving the characters of H. In particular, if we are given C and we let H be the cyclic subgroup generated by g_C then we are able to express $\tilde{\pi}(x, \delta_C)$ in terms of Hecke L-functions. As we know the analytic properties of these L-functions, we are now able to follow rather closely the classical method as developed in [D] to prove Theorem 7.1.

Though the above technique has the advantage of replacing the non-abelian L-functions with abelian ones, it does so at some cost. The estimates will now involve the field constants (that is, degree, discriminant, etc.) of the fixed field M (say) of H. In general, as we do not have any information about M we are forced to majorize its field constants by those of K and this magnifies the error terms significantly.

This problem could be avoided if we were able to deal directly with the Artin L-functions. This theme is developed in the next section. We conclude this section by stating some unconditional results developed in [LO] and in [LMO].

Theorem 7.2 *If* $\log x \gg n_L (\log d_L)^2$, *then*

$$|\pi_C(x) - \frac{|C|}{|G|} \operatorname{Li}(x)| \le \frac{|C|}{|G|} \operatorname{Li}(x^\beta) + \mathbf{O}(|\tilde{C}|x \exp(-cn_L^{-\frac{1}{2}}(\log x)^{\frac{1}{2}}))$$

where $|\tilde{C}|$ *is the number of conjugacy classes contained in* C, β *is the exceptional zero of Proposition 6.1, and the term* $\frac{|C|}{|G|} \operatorname{Li}(x^\beta)$ *is to be suppressed if the exceptional zero* β *does not exist.*

Sometimes it is useful to have an inequality rather than an explicit error term. Such a bound is provided by the following result of Lagarias, Odlyzko and Montgomery [LMO].

Theorem 7.3 *We have*

$$\pi_C(x) \ll \frac{|C|}{|G|} \operatorname{Li}(x)$$

provided

$$\log x \gg (\log d_L)(\log \log d_L)(\log \log \log e^{20} d_L).$$

In applying these results, it is very useful to have some estimates for the discriminant of a field. These upper bounds are consequences of an inequality due to Hensel, and are developed in [Se2]. Let $\mathfrak{D}_{K/F}$ denote the different of K/F. It is an ideal of $\mathcal{O}_K$ and its norm $\mathfrak{d}_{K/F}$ from K to F is the discriminant of the extension. Let ν be a place of F and w a place of K dividing it. Let p_ν denote the residue characteristic of ν. Hensel's estimate states

$$w(\mathfrak{D}_{K/F}) = e_{w/\nu} - 1 + s_{w/\nu}$$

where

$$0 \le s_{w/\nu} \le w(e_{w/\nu}).$$

Here $e_{w/\nu}$ is the ramification index of p_ν in K. Using this, one can get an estimate for the norm of the relative discriminant. Let us set

$$n_K = [K : \mathbb{Q}], \quad n_F = [F : \mathbb{Q}]$$

and

$$n = [K : F] = n_K / n_F.$$

Let us also set $P(K/F)$ to be the set of rational primes p for which there is a prime $\mathfrak{p}$ of F with $\mathfrak{p}|p$ and $\mathfrak{p}$ is ramified in K. Then,

$$\log \mathbf{N}_{F/\mathbb{Q}} \mathfrak{d}_{K/F} \le (n_K - n_F) \sum_{p \in P(K/F)} \log p + n_K (\log n)|P(K/F)|.$$

This bound does not assume that K/F is Galois. If we know in addition that K/F is Galois, the following slightly stronger estimate holds:

$$\log \mathbf{N}_{F/\mathbb{Q}} \mathfrak{d}_{K/F} \le (n_K - n_F) \sum_{p \in P(K/F)} \log p + n_K (\log n).$$

There is an analogue of this for Artin conductors also. This analogue is needed in the proofs of the results of the next section.

Proposition 7.4 *Suppose that K/F is Galois with group G. Let χ denote an irreducible character of G and denote by $\mathfrak{f}_\chi$ its Artin conductor. Then*

$$\log \mathbf{N}_{F/\mathbb{Q}}\mathfrak{f}_\chi \leq 2\chi(1)n_F\{ \sum_{p\in P(K/F)} \log p + \log n\}.$$

Proof. Firstly, we observe that for each $i \geq 0$,

$$\dim V^{G_i} = \frac{1}{|G_i|} \sum_{a\in G_i} \chi(a),$$

where G_i is as in Section 2.

Thus, for each finite ν,

$$n(\chi,\nu) = \sum_i \frac{|G_i|}{|G_0|}\left(\chi(1) - \frac{1}{|G_i|}\sum_{a\in G_i}\chi(a)\right).$$

Denote by $\mathcal{O}_\nu$ (respectively $\mathcal{O}_w$) the ring of integers of F_ν (resp. K_w). Define a function i_G on G by

$$i_G(g) = w(gx - x) = \max\{i : g \in G_{i-1}\}$$

where $\mathcal{O}_w = \mathcal{O}_\nu[x]$. Rearranging gives

$$n(\chi,\nu) = \frac{\chi(1)}{|G_0|}\sum_i(|G_i| - 1) - \frac{1}{|G_0|}\sum_{1\neq a\in G_0}\chi(a)i_G(a).$$

Applying this formula for χ the trivial character, and the character of the regular representation of G_0, we find that

$$\sum_{1\neq a\in G_0} i_G(a) = \sum_i(|G_i| - 1) = w(\mathfrak{D}_{K/F}).$$

Hence,

$$n(\chi,\nu) = \frac{1}{|G_0|}\sum_{1\neq a\in G_0} i_G(a)(\chi(1) - \chi(a)) \leq \frac{2\chi(1)w(\mathfrak{D}_{K/F})}{e_{w/\nu}}.$$

Now using the above stated estimate for $w(\mathfrak{D}_{K/F})$ we deduce that

$$\log \mathbf{N}\mathfrak{f}_\chi \leq 2\chi(1)\sum \frac{1}{e_{w/\nu}}(e_{w/\nu} - 1 + s_{w/\nu})f_\nu \log p_\nu$$

and this is

$$\leq 2\chi(1) \left\{ \sum f_\nu \left(1 - \frac{e_\nu}{e_w}\right) \log p_\nu + \sum f_\nu \frac{e_\nu}{e_w} w(e_{w/\nu}) \log p_\nu \right\}$$

where e_ν (resp. e_w) denotes absolute ramification index at ν (resp. w) and we have used $e_{w/\nu} = e_w/e_\nu$. Also, as $w(e_{w/\nu}) = e_w \nu_p(e_{w/\nu})$ and as K/F is Galois, $e_{w/\nu}$ divides n. Thus

$$\log \mathrm{N}\mathfrak{f}_\chi \leq 2\chi(1) n_F \left\{ \sum_{p \in P(K/F)} \log p + \log n \right\}.$$

This completes the proof.

We remark that there is no analogue of Hensel's estimate in the function field case. This is one source of difficulty in extending to this case the effective versions of the Chebotarev density theorem discussed in this and the next section. The reader is referred to [MS] and the references therein for the function field analogues.

§8 Consequences of Artin's conjecture

These estimates can be significantly improved if we know Artin's conjecture on the holomorphy of L-series. The improvement is in the dependence of the error term on C. The results of this section are from the paper [MMS]. We shall only discuss the conditional result Proposition 7.1.

Let χ be a character of G and denote by $\pi(x, \chi)$ the function

$$\pi(x, \chi) = \sum_{\mathrm{N}\nu \leq x} \chi(\sigma_\nu).$$

Let $\delta(\chi)$ denote the multiplicity of the trivial character in χ.

As before (see §2)

$$A_\chi = d_K^{\chi(1)} \mathrm{N}_{F/\mathbb{Q}}(\mathfrak{f}_\chi)$$

and

$$\Lambda(s, \chi) = \Lambda(s, \chi, F) = A_\chi^{s/2} \gamma(s, \chi, F) L(s, \chi)$$

Proposition 8.1 *Suppose that the Artin L-series $L(s, \chi)$ is analytic for all $s \neq 1$ and is nonzero for $\mathrm{Re}(s) \neq \frac{1}{2}, 0 < \mathrm{Re}(s) < 1$. Then*

$$\pi(x, \chi) = \delta(\chi) \mathrm{Li}(x) + \mathbf{O}(x^{\frac{1}{2}}((\log A_\chi) + \chi(1) n_F \log x)) + \mathbf{O}(\chi(1) n_F \log M(K/F))$$

where

$$M(K/F) = n d_F^{1/n_F} \prod_{p \in P(K/F)} p.$$

Proof. The argument proceeds along standard lines and so we just sketch it here. Artin proved the functional equation

$$\Lambda(s,\chi) = W(\chi)\Lambda(1-s,\bar\chi)$$

where $W(\chi)$ is a complex number of absolute value 1 and $\bar\chi$ is the complex conjugate of χ. We know that

$$(s(s-1))^{\delta(\chi)}\Lambda(s,\chi)$$

is entire and we have the Hadamard factorization

$$\Lambda(s,\chi) = e^{a(\chi)+b(\chi)s}\prod_\rho(1-\frac{s}{\rho})e^{s/\rho}(s(s-1))^{-\delta(\chi)}$$

where $a(\chi), b(\chi) \in \mathbb{C}$ and the product runs over all zeroes ρ of $\Lambda(s,\chi)$ (necessarily $0 \le \operatorname{Re}(\rho) \le 1$.) From the equality

$$\overline{\Lambda(s,\chi)} = \Lambda(\bar s, \bar\chi)$$

we deduce the relation

$$\overline{\frac{\Lambda'}{\Lambda}(s,\chi)} = \frac{\Lambda'}{\Lambda}(\bar s,\bar\chi).$$

Moreover, the functional equation implies the relation

$$\frac{\Lambda'}{\Lambda}(s,\chi) = -\frac{\Lambda'}{\Lambda}(1-s,\bar\chi).$$

From these two relations, we deduce that

$$\operatorname{Re}\frac{\Lambda'}{\Lambda}(\frac{1}{2},\chi) = 0.$$

Also, if ρ is a zero of $\Lambda(s,\chi)$ then so is $1-\bar\rho$. Hence,

$$\operatorname{Re}\sum(\frac{1}{2}-\rho)^{-1} = 0$$

as is seen by grouping together the terms corresponding to ρ and $1-\bar\rho$ in the absolutely convergent sum. Logarithmically differentiating the product formula at $s = \frac{1}{2}$ and taking real parts, we deduce that

$$\operatorname{Re}(b(\chi) + \sum_\rho\frac{1}{\rho}) = 0.$$

Hence,

$$\operatorname{Re}\frac{\Lambda'}{\Lambda}(s,\chi) = \sum_\rho\operatorname{Re}(\frac{1}{s-\rho}) - \delta(\chi)\operatorname{Re}(\frac{1}{s}+\frac{1}{s-1}).$$

Let $N(t, \chi)$ denote the number of zeros $\rho = \beta + i\gamma$ of $L(s, \chi)$ with $0 < \beta < 1$ and $|\gamma - t| \leq 1$. Evaluating the above formula at $s = 2 + it$ and observing that

$$\text{Re}(\frac{1}{2 + it - \rho}) = \frac{2 - \beta}{(2 - \beta)^2 + (t - \gamma)^2}$$

is non-negative for all ρ and is atleast $1/5$ if $|t - \gamma| \leq 1$ we deduce that

$$N(t, \chi) \ll \text{Re}\, \frac{\Lambda'}{\Lambda}(2 + it, \chi).$$

Since the Dirichlet series for $L(s, \chi)$ converges at $2 + it$, the right hand side is easily estimated, the essential contribution coming from $\log A_\chi$ and the number of Γ factors. We get

$$N(t, \chi) \ll \log A_\chi + \chi(1) n_F \log(|t| + 5).$$

By developing an explicit formula as in [LO] or [Mu2] we find that

$$\sum_{\mathbf{N}\nu \leq x}{}' \chi(\sigma_\nu) \log \mathbf{N}\nu = \delta(\chi)x - \sum_{|\gamma| < x} \frac{x^\rho}{\rho} + \mathbf{O}(\chi(1) n_F \log M(K/F))$$
$$+ \mathbf{O}(x^{\frac{1}{2}}(\log x)(\log A_\chi + \chi(1) n_F \log x)),$$

where the prime on the sum indicates that we only include places ν that are unramified in K. The sum over zeros can be estimated by observing that

$$\sum_{|\gamma| < x} \frac{1}{\rho} \ll \sum_{j < x} \frac{N(j, \chi)}{j}$$

and using the above estimate for $N(t, \chi)$. The estimate for $\pi(x, \chi)$ can be deduced by partial summation.

Proposition 8.2 *Suppose that all Artin L-series of the extension K/F are analytic at $s \neq 1$ and that GRH holds. Then*

$$\sum_C \frac{1}{|C|} \left(\pi_C(x) - \frac{|C|}{|G|} \text{Li}\, x \right)^2 \ll x n_F^2 (\log M(K/F) x)^2.$$

Proof. We first observe that

$$\sum_C \frac{1}{|C|} \left(\frac{|C|}{|G|} \pi(x, 1_G) - \frac{|C|}{|G|} \text{Li}\, x \right)^2 = \frac{1}{|G|}(\pi(x, 1_G) - \text{Li}\, x)^2.$$

Expressing $\pi(x, 1_G)$ in terms of characters, we see that this is

$$\leq \frac{1}{|G|} \left(\sum_{\chi \neq 1} |\pi(x, \chi)|^2 + (\pi(x, 1_G) - \operatorname{Li} x)^2 \right)$$

where the sum is over the non-trivial irreducible characters of G. By Propositions 7.4 and 8.1,

$$\pi(x, \chi) - \delta(\chi) \operatorname{Li} x \ll \chi(1) n_F x^{\frac{1}{2}} (\log(M(K/F)x)).$$

The result follows on noting that

$$\sum_{\chi} \chi(1)^2 = |G|.$$

Proposition 8.3 *Let D be a union of conjugacy classes. Under the same hypotheses as in Proposition 8.2, we have*

$$\pi_D(x) = \frac{|D|}{|G|} \operatorname{Li} x + \mathbf{O}(|D|^{\frac{1}{2}} x^{\frac{1}{2}} n_F \log M(K/F)x).$$

Proof. We have

$$\pi_D(x) - \frac{|D|}{|G|} \operatorname{Li} x = \sum_{C} \left(\pi_C(x) - \frac{|C|}{|G|} \operatorname{Li} x \right)$$

where the sum is taken over all conjugacy classes C contained in D. Now applying the Cauchy-Schwarz inequality we deduce that

$$\sum_{C} \left| \pi_C(x) - \frac{|C|}{|G|} \operatorname{Li} x \right| \ll \left(\sum_{C} |C| \right)^{\frac{1}{2}} \left(\sum_{C} \frac{1}{|C|} \left| \pi_C(x) - \frac{|C|}{|G|} \operatorname{Li} x \right|^2 \right)^{\frac{1}{2}}.$$

The result now follows from Proposition 8.2.

Remark. Using Hensel's estimate for the discriminant, it is possible to write the error term in Theorem 7.1 as

$$\mathbf{O}(|C| x^{\frac{1}{2}} n_F \log M(K/F)x).$$

Thus Artin's conjecture allows us to replace $|C|$ with $|C|^{\frac{1}{2}}$. In some cases, we can also improve Theorem 7.1 even without assuming Artin's conjecture. We give two such results below.

Proposition 8.4   *Let D be a union of conjugacy classes in G and let H be a subgroup of G satisfying*
(i)   *Artin's conjecture is true for the irreducible characters of H*
(ii)   *H meets every class in D.*
    Suppose the GRH holds. Then

$$\pi_D(x) = \frac{|D|}{|G|} \operatorname{Li} x + \mathbf{O}\left(x^{\frac{1}{2}} \left(\sum_{C \subseteq D} \frac{|C|^2}{|C_H|} \right)^{\frac{1}{2}} n_F \log Mx \right)$$

where $M = M(K/F)$ and $C_H = C_H(\gamma)$ for some $\gamma \in H \cap C$.

Proof. Firstly, we have the relation

$$\pi_D(x) = \tilde{\pi}_D(x) + \mathbf{O}(\frac{1}{|G|} \log d_K + n_F x^{\frac{1}{2}}).$$

Using the estimate from Hensel's bound, we have

$$\frac{1}{|G|} \log d_K \ll n_F \log Mx.$$

Also,

$$\tilde{\pi}_D(x) = \sum_{C \subseteq D} \tilde{\pi}_C(x) = \sum_{C \subseteq D} \frac{|C|}{|G|} \cdot \frac{|H|}{|C_H|} \tilde{\pi}_{C_H}(x). \tag{8.1}$$

Now,

$$\sum_{C \subseteq D} \frac{|C|}{|G|} \cdot \frac{|H|}{|C_H|} (\tilde{\pi}_{C_H}(x) - \pi_{C_H}(x))$$

$$\leq \frac{|H|}{|G|} \sum_{C \subseteq D} \frac{|C|}{|C_H|} \left(\sum_{\substack{N\nu^m \leq x \\ m \geq 2}} \delta_{C_H}(\sigma_\nu^m) + \sum_{\substack{N\nu \leq x \\ \nu \text{ ramified in } L/K}} \delta_{C_H}(\sigma_\nu) \right)$$

$$\leq \frac{|H|}{|G|} \left(\max_{C \subseteq D} \frac{|C|}{|C_H|} \right) \cdot \left\{ \frac{|G|}{|H|} n_F x^{\frac{1}{2}} + \frac{2}{\log 2} \frac{1}{|H|} \log d_K \right\}$$

$$\leq \left(\max \frac{|C|}{|C_H|} \right) (n_F x^{\frac{1}{2}} + n_F \log Mx)$$

and this can be absorbed into the error term. Therefore, we can replace $\tilde{\pi}_{C_H}$ by π_{C_H} in the equation (8.1). Now,

$$\sum_{C \subseteq D} \frac{|C|}{|G|} \cdot \frac{|H|}{|C_H|} \pi_{C_H}(x)$$

$$= \frac{|D|}{|G|} \operatorname{Li} x + \mathbf{O}\left(\frac{|H|}{|G|} \cdot \sum_{C \subseteq D} \frac{|C|}{|C_H|^{\frac{1}{2}}} \frac{1}{|C_H|^{\frac{1}{2}}} \left| \pi_{C_H}(x) - \frac{|C_H|}{|H|} \operatorname{Li} x \right| \right).$$

Now applying the Cauchy-Schwarz inequality and using Proposition 8.2, we find that the error term above is

$$\ll \frac{|H|}{|G|} \cdot \left(\sum_{C \subseteq D} \frac{|C|^2}{|C_H|} \right)^{\frac{1}{2}} x^{\frac{1}{2}} n_F \frac{|G|}{|H|} \log M(K/F')x$$

where F' is the fixed field of H. This proves the proposition since $M(K/F') \ll M(K/F)$.

We state one immediate corollary of this result.

Corollary 8.5   *Under the same hypotheses as above,*

$$\pi_D(x) = \frac{|D|}{|G|} \operatorname{Li} x + \mathbf{O}\left(|D|^{\frac{1}{2}} x^{\frac{1}{2}} \left(\frac{|G|}{|H|} \right)^{\frac{1}{2}} n_F \log Mx \right).$$

The corollary follows immediately on noting that

$$\frac{|C|}{|C_H|} \leq \frac{|G|}{|H|}.$$

We now present one further result in this direction. This estimate has the feature that in some cases, it gives a better result than what one deduces from Artin's conjecture.

Proposition 8.6   *Suppose the GRH holds. Let D be a nonempty union of conjugacy classes in G and let H be a normal subgroup of G such that Artin's conjecture is true for the irreducible characters of G/H and $HD \subseteq D$. Then*

$$\pi_D(x) = \frac{|D|}{|G|} \operatorname{Li} x + \mathbf{O}\left(\left(\frac{|D|}{|H|} \right)^{\frac{1}{2}} x^{\frac{1}{2}} n_F \log Mx \right)$$

where M is as in the previous proposition.

Proof. Let $\bar{D}$ be the image of D in G/H. It is a union of conjugacy classes in G/H and

$$\pi_{\bar{D}}(x) = \frac{|\bar{D}| \cdot |H|}{|G|} \operatorname{Li} x + \mathbf{O}(|\bar{D}|^{\frac{1}{2}} x^{\frac{1}{2}} n_F \log M(F'/F)x)$$

where F' is the fixed field of H. As $HD \subseteq D$,

$$|\bar{D}| \cdot |H| = |D|$$

and

$$\pi_D(x) = \pi_{\bar{D}}(x) + \mathbf{O}((\log d_K)/|G|).$$

Also, $M(F'/F) \ll M(K/F)$. The result follows.

Finally, we can ask what the true order of the error term in the Chebotarev theorem should be. Let $\alpha(G)$ denote the number of conjugacy classes of G.
Question. Is it true that for any conjugacy set $D \subseteq G$,

$$\pi_D(x) = \frac{|D|}{|G|} \operatorname{Li} x + \mathbf{O}\left(\left(\frac{|D|}{\alpha(G)} \right)^{\frac{1}{2}} x^{\frac{1}{2}} n_F \log M x \right)?$$

This would be implied by the Proposition 8.2 if all the terms are of the same order. In the case $F = \mathbb{Q}$ and K/F is Abelian, our question is a well-known conjecture of Montgomery.

§9 The least prime in a conjugacy class

Let L/K be a finite non-trivial Galois extension of number fields with group G. Our main result is an estimate, assuming the Riemann Hypothesis for Dedekind zeta functions (GRH), for the least norm of a prime ideal of K which is unramified in L and which does not split completely. The results of this section are from [Mu3].

If C is any subset of G stable under conjugation, Lagarias and Odlyzko [LO, pp. 461–462] showed, assuming (GRH) that there is a prime ideal $\mathfrak{p}$ with

$$N_{K/\mathbb{Q}}\mathfrak{p} \ll (\log |d_L|)^2 \tag{9.1}$$

for which the Frobenius conjugacy class $\sigma_{\mathfrak{p}}$ of $\mathfrak{p}$ lies in C. Here, d_L (resp. d_K) denotes the (absolute) discriminant of L (resp. K). In this estimate, an important tool was the effective version of the Chebotarev density theorem proved in [LO]. By the results of the previous section, it follows that the assumption of Artin's conjecture (AC) on the holomorphy of Artin L- series allows one to prove a sharper version of this theorem. In particular, the assumption of AC implies that the estimate (9.1) may be improved to

$$\mathbf{N}_{K/\mathbb{Q}}\mathfrak{p} \ll \frac{(\log |d_L|)^2 (\log |G|)^2}{|C|}. \tag{9.2}$$

In fact, the term $(\log |G|)^2$ may also be removed by using a more detailed argument. The purpose of this section is to show, assuming the GRH, that in the special case $C = G - \{1\}$, there is a prime ideal $\mathfrak{p}$ of K of degree 1 which is unramified in L which does not split completely and which satisfies

$$\mathbf{N}_{K/\mathbb{Q}}\mathfrak{p} \ll \left(\frac{\log |d_L|}{|G| - 1} \right)^2 \ll \left(\frac{n_K}{n_L} \log |d_L| \right)^2. \tag{9.3}$$

where $n_K = [K : \mathbb{Q}]$ and $n_L = [L : \mathbb{Q}]$. Thus, the estimate (9.3) shows that for the special set $C = G - \{1\}$, one can do substantially better than (9.1).

Next, we shall show that for certain subgroups H of G the bound (9.3) may be extended for the least norm of a prime ideal $\mathfrak{p}$ for which $\sigma_{\mathfrak{p}}$ does not intersect H. The precise statement is given in Theorem 9.3.

We apply these last results to the group of points on an elliptic curve over a finite field. Let E be an elliptic curve without complex multiplication and defined over $\mathbb{Q}$. Denote by $\mathcal{N}$ the conductor of E.

Let us set
$$T = \mathrm{lcm}_{E'}|E'(\mathbb{Q})_{\mathrm{tors}}|$$
where the lcm ranges over elliptic curves E' which are defined over $\mathbb{Q}$ and are $\mathbb{Q}$-isogenous to E. In [K, Th. 2] Katz proved that
$$\gcd |E(\mathbb{F}_p)| = T$$
where the *gcd* is taken over primes p of good reduction. It is well known and easily proved that both sides are divisible by the same primes. Using our results, we can make this effective in the following sense. Let $l \geq 5$ be a prime and assume the *GRH*. If l does not divide T then we show (Theorem 9.4) that there is a prime p so that
$$p \ll (\ell \log \mathcal{N}\ell)^2$$
and $E(\mathbb{F}_p)$ does not have a point of order ℓ.

We begin by proving the estimate (9.3). We recall that L/K is a *non-trivial* Galois extension.

Theorem 9.1 *Assume the GRH. Then, there exists a prime ideal $\mathfrak{p}$ of K*
(i) *$\mathfrak{p}$ is of degree 1 over $\mathbb{Q}$ and unramified in L*
(ii) *$\mathfrak{p}$ does not split completely in L*
and
$$\mathbf{N}_{K/\mathbb{Q}}\mathfrak{p} \ll (\frac{n_K}{n_L} \log |d_L|)^2$$
where $n_K = [K : \mathbb{Q}]$ and $n_L = [L : \mathbb{Q}]$.

Proof. We consider the kernel function of [LMO, §2], namely
$$k(s) = k(s; x, y) = \left(\frac{y^{s-1} - x^{s-1}}{s-1} \right)^2.$$

For $y > x > 1$ and $u > 0$, it has the property that the inverse Mellin transform
$$\hat{k}(u) = \frac{1}{2\pi i} \int_{(2)} k(s)u^{-s}ds$$
is given by the formulae
$$\hat{k}(u; x, y) = \begin{cases} 0 & \text{if} \quad u > y^2 \\ \frac{1}{y} \log \frac{y^2}{u} & \text{if} \quad xy < u < y^2 \\ \frac{1}{u} \log \frac{u}{x^2} & \text{if} \quad x^2 < u < xy \\ 0 & \text{if} \quad u < x^2. \end{cases}$$

Now consider the integral

$$J_K = \frac{1}{2\pi i} \int_{(2)} \left(-\frac{\zeta_K'}{\zeta_K}(s) \right) k(s; x, y) ds.$$

On the one hand, it is equal to

$$(\log y/x)^2 - \sum_\rho k(\rho; x, y)$$

where ρ runs over all zeroes of $\zeta_K(s)$. Write $\rho = \beta + i\gamma$. If $N_K(r; s_0)$ denotes the number of zeroes ρ of $\zeta_K(s)$ with $|\rho - s_0| \le r$ then ([LMO, Lemma 2.2])

$$N_K(r; s_0) \ll 1 + r(\log |d_K| + n_K \log(|s_0| + 2)).$$

Since

$$|k(\rho; x, y)| \le \frac{x^{-2(1-\beta)}}{|\rho - 1|^2}$$

it follows that

$$\sum_{\beta \le 1-\delta} k(\rho; x, y) \ll x^{-2\delta} \int_\delta^\infty \frac{1}{r^2} dN_K(r; 1)$$

$$\ll x^{-2\delta}(\delta^{-2} + \delta^{-1} \log |d_K|).$$

As we are assuming the GRH, we may take $\delta = \frac{1}{2}$ and we see that

$$J_K = (\log \frac{y}{x})^2 + \mathbf{O}(x^{-1} \log |d_K|).$$

On the other hand, the integral is equal to the sum

$$\sum_{\mathfrak{p}, m} \Lambda((N\mathfrak{p})^n) \hat{k}((N\mathfrak{p})^n; x, y).$$

The contribution to this sum of ideals $\mathfrak{p}^n$ for which $N\mathfrak{p}^n$ is not a rational prime is

$$\ll \frac{n_K (\log y)(\log y/x)}{x \log x}$$

as in [LMO, (2.6)]. Moreover, the contribution of primes $\mathfrak{p}$ which ramify in L is

$$\ll \sum_{\mathfrak{p} | d_{L/K}} (\log N\mathfrak{p}) x^{-2} \log \frac{y}{x}$$

as in [LMO, (2.27)]. (Recall that $d_{L/K}$ is the norm to $\mathbb{Q}$ of the discriminant of the extension L/K.) Since [Se2, p. 129]

$$\sum_{\mathfrak{p} | d_{L/K}} \log N\mathfrak{p} \le \frac{2}{n} \log |d_L|,$$

the contribution of primes $\mathfrak{p}$ which ramify in L is

$$\ll \frac{1}{n}(\log |d_L|)x^{-2}\log\frac{y}{x}.$$

Let us set

$$\tilde{J}_K = \sum{}^* (\log N\mathfrak{p})\hat{k}(N\mathfrak{p}; x, y)$$

where the sum ranges over primes $\mathfrak{p}$ of K of degree 1 which are unramified in L. Then the above estimates imply that

$$\tilde{J}_K = (\log\frac{y}{x})^2 + \mathbf{O}\left(x^{-1}\log|d_K| + n_K(\log y)(\log\frac{y}{x})\frac{1}{x\log x} + (\frac{1}{n}\log|d_L|)x^{-2}(\log\frac{y}{x})\right).$$

On the other hand, by an argument similar to that given above,

$$J_L = (\log\frac{y}{x})^2 + \mathbf{O}(x^{-1}\log|d_L|).$$

Now if we suppose that *every* prime ideal $\mathfrak{p}$ of K with $N\mathfrak{p} \le y^2$ either ramifies or splits completely in L, then

$$J_L \ge n\tilde{J}_K.$$

Putting this together with the above estimates, and choosing

$$x = (\frac{\alpha}{n}\log|d_L|)$$

and

$$y = bx$$

for some $b > 1$ and $\alpha > (\log b)^2$ we deduce the inequality

$$(n-1)(\log b)^2 \ll \frac{n}{\alpha} + \frac{nn_L(\log bx)(\log b)}{\alpha(\log|d_L|)(\log x)} + \frac{n^2(\log b)}{\alpha^2(\log|d_L|)} \ll n.$$

For a sufficiently large value of b, we get a contradiction. This completes the proof.

Remarks. 1. This method can also be used to produce an unconditional bound. In terms of its dependence on L the main term is $|d_L|^{1/2(n-1)}$.

2. Note that we used the normality of the extension L/K in asserting that a prime of K which splits completely in L has $[L:K]$ prime divisors in L.

3. We note an interesting consequence of the above. Assume the GRH. Suppose the class number h of K is larger than 1. There exists a non-principal prime ideal $\mathfrak{p}$ of K of degree 1 over $\mathbb{Q}$ with

$$N_{K/\mathbb{Q}}\mathfrak{p} \ll (\log|d_K|)^2.$$

Indeed, choose for L the Hilbert class field of K, and use the fact that $d_L = d_K^h$.

We describe two variants of Theorem 9.1.

(A) Consider the following diagram of fields.

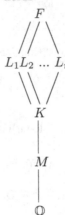

Theorem 9.2   *Assume the GRH. Let $L_1, \ldots, L_r$ be distinct non-trivial Galois extensions of K. Let F be an extension of K containing all the L_i and M a subfield of K so that F/M is Galois. Set*

$$m = \min[L_i : K]$$
$$f = [F : K]$$

and assume that

$$r < m.$$

Then, there exists a prime ideal $\mathfrak{p}$ of K satisfying
(i') $\mathfrak{p}$ is of degree 1 over $\mathbb{Q}$ and $N_{K/M}\mathfrak{p}$ does not ramify in F
(ii') $\mathfrak{p}$ does not split completely in any of the L_i, $1 \leq 1 \leq r$
    with

$$N_{K/\mathbb{Q}}\mathfrak{p} \ll B^2$$

    where
$$B = \max\left(\frac{\sum_{i=1}^r (\log |d_{L_i}|)}{m - r}, \sqrt{\frac{m}{f(m-r)} \log |d_F|} \right).$$

Proof. Let $\mathcal{S}$ denote the set of degree 1 prime ideals $\mathfrak{p}$ of K with $N\mathfrak{p} \leq y^2$ for which $\mathfrak{p} \cap \mathcal{O}_M$ does not ramify in F. Suppose that every element of $\mathcal{S}$ splits completely in *some* L_i. Then, with notation as in the proof of Theorem 9.1, we have

$$\sum J_{L_i} \geq m \sum_{\mathfrak{p} \in \mathcal{S}} (\log N\mathfrak{p})\hat{k}(N\mathfrak{p}; x, y).$$

Using the estimate for J_L and $\tilde{J}_K$ given in the proof of Theorem 9.1, we deduce that

$$r(\log \frac{y}{x})^2 + \mathbf{O}(\frac{1}{x} \sum_i \log |d_{L_i}|)$$

$$\geq m(\log \frac{y}{x})^2 + \mathbf{O}(\frac{m}{x} \log |d_K|) + \mathbf{O}(\frac{mn_K (\log y)(\log y/x)}{x \log x})$$

$$+ \mathbf{O}(\frac{m}{f}(\log |d_F|)\frac{1}{x^2} \log y/x).$$

Simplifying, and choosing $x = \alpha B$ and $y = \beta x$ with some $\beta > 1$ and $\alpha > (\log \beta)^2$, we get the inequality

$$(m - r)(\log \beta)^2 \leq \mathbf{O}((m - r)(\log \beta))$$

which is a contradiction if β is sufficiently large.

(B) With L/K a normal extension and $G = \text{Gal}(L/K)$ as before, we take a subgroup H of G. We want to find a prime $\mathfrak{p}$ of K so that $\sigma_{\mathfrak{p}}$ is disjoint from H. Theorem 9.1 had to do with $H = \{1\}$.

Theorem 9.3. *Assume the GRH. Denote by $N = N_G(H)$ the normalizer of H in G and let R be the fixed field of N. Let $H_1, \ldots, H_r$ be a set of normal subgroups of N and $L_1, \ldots, L_r$ their respective fixed fields. Suppose that*
(1) *for each $g \in G$, $gHg^{-1} \cap N$ is contained in some $H_i (1 \leq i \leq r)$.*
(2) *if $m = \min[L_i : R]$ then $r < m$.*
    Then, there exists a prime ideal $\mathfrak{p}$ of K with

$$N_{K/\mathbb{Q}}\mathfrak{p} \ll B_H^2$$

    and satisfying
(a) $\mathfrak{p}$ *is of degree one and does not ramify in L*
(b) $\sigma_{\mathfrak{p}}$ *is disjoint from H.*
    Here,

$$B_H = \max \left\{ \frac{1}{m-r} \left(\sum \frac{1}{|H_i|} \right) \log |d_L|, \sqrt{\frac{m}{|N|(m-r)}} (\log |d_L|) \right\}.$$

Proof. Each L_i is a Galois extension of R and L is a Galois extension of R containing all the L_i. By Theorem 9.2, we can find a prime ideal $\mathfrak{P}$ of R of degree one (over $\mathbb{Q}$) so that $\mathfrak{p} = N_{R/K}\mathfrak{P}$ does not ramify in L, $\mathfrak{P}$ does not split completely in any of the L_i and

$$N_{R/\mathbb{Q}}\mathfrak{P} \ll B^2$$

where

$$B = \max \left\{ \frac{1}{m-r} \sum_{i=1}^r \log |d_{L_i}|, \sqrt{\frac{m}{[L:R](m-r)}} (\log |d_L|) \right\}.$$

The splitting completely condition means that

$$\sigma_{\mathfrak{P}} \cap H_i = \phi \qquad 1 \leq i \leq r.$$

Now

$$\sigma_{\mathfrak{p}} = \cup_\tau \tau \sigma_{\mathfrak{P}} \tau^{-1}$$

where the union is over a set of coset representatives $\{\tau\}$ for N in G. It follows that

$$\sigma_{\mathfrak{p}} \cap H = \phi.$$

Hence $\mathfrak{p}$ satisfies (a) and (b). Now as

$$|d_{L_i}| \leq |d_L|^{1/|H_i|},$$

we deduce the stated bound.

Remarks. 1. In the case $r = 1$, the assumptions (1) and (2) may be stated as
(1') for any $g \in G$, $gHg^{-1} \cap N$ is nonempty $\Rightarrow g \in N$
(2') H is a proper subgroup of N.
2. If we are only interested in finding a prime $\mathfrak{p}$ so that $\sigma_{\mathfrak{p}} = (\mathfrak{p}, L/K)$ is not *contained* in H, then we do not need to consider the conjugates of H at all. Rather, it suffices to take a degree one prime $\mathfrak{P}$ of R such that $\mathfrak{p} = \mathfrak{P} \cap \mathcal{O}_K$ does not ramify in L and $(\mathfrak{P}, L/R)$ is not contained in H. But as H is normal in N, this just means that $\mathfrak{P}$ does not split completely in $M = $ the fixed field of H. We can find such a $\mathfrak{P}$ with

$$N_{R/\mathbb{Q}}\mathfrak{P} \ll \left(\frac{\log|d_L|}{|N| - |H|} \right)^2.$$

Corollary 9.4. *Let the notation and hypotheses be as in Theorem 9.3. If C is a subset of G stable under conjugation and H intersects every conjugacy class in C nontrivially, then there is a prime $\mathfrak{p}$ of K satisfying*

$$N_{K/\mathbb{Q}}\mathfrak{p} \ll B_H^2$$

as well as (a) *and*
(b') $\sigma_{\mathfrak{p}}$ *is not contained in C.*

Let E be an elliptic curve defined over $\mathbb{Q}$ and let $\mathcal{N}$ denote its conductor. For $p \nmid \mathcal{N}$, we may consider the group $|E(\mathbb{F}_p)|$ of $\mathbb{F}_p$-rational points on E. Its cardinality is given by $|E(\mathbb{F}_p)| = p + 1 - a(p)$ for some integer $a(p)$.

The action of $\mathrm{Gal}(\bar{\mathbb{Q}}/\mathbb{Q})$ on points of $E(\bar{\mathbb{Q}})$ which are in the kernel of multiplication by ℓ gives a representation

$$\rho_\ell : \mathrm{Gal}(\bar{\mathbb{Q}}/\mathbb{Q}) \to GL_2(\mathbb{F}_\ell).$$

It has the property that for $p \nmid \ell\mathcal{N}$, $\rho_\ell(\sigma_p)$ has trace $a(p)$ and determinant p modulo ℓ.

Recall that we have set

$$T = \mathrm{lcm}_{E'}|E'(\mathbb{Q})_{\mathrm{tors}}|$$

where the lcm ranges over elliptic curves E' which are $\mathbb{Q}$-isogenous to E.

Theorem 9.5 *Suppose that E does not have complex multiplication and let $\ell \geq 5$ be a prime which does not divide T. Denote by $\mathcal{N}$ the conductor of E. Assume the GRH. Then, there is a prime*

$$p \ll (\ell \log \ell \mathcal{N})^2$$

such that $E(\mathbb{F}_p)$ does not have a point of order ℓ.

Proof. Let us denote by G the image of ρ_ℓ. It is known that the fixed field of the kernel of ρ_ℓ contains the field of ℓ-th roots of unity. Let PG denote the image of G under the natural map $GL_2(\mathbb{F}_\ell) \to PGL_2(\mathbb{F}_\ell)$. It is well known (See [Se2, p. 197]) that one of the following holds:

(i) PG contains $PSL_2(\mathbb{F}_\ell)$
(ii) G is contained in a Borel subgroup of $GL_2(\mathbb{F}_\ell)$
(iii) G is contained in a non-split Cartan subgroup of $GL_2(\mathbb{F}_\ell)$
(iv) $PG \simeq A_4$, S_4 or A_5
(v) G is contained in the normalizer of a Cartan subgroup C but is not contained in C.

We shall consider each in turn.

(i): Consider the Borel subgroup (see [Se2, p. 197])

$$B = \left\{ \begin{pmatrix} * & * \\ 0 & * \end{pmatrix} \right\} \subseteq G = GL_2(\mathbb{F}_\ell)$$

and the subgroups

$$H = \left\{ \begin{pmatrix} 1 & * \\ 0 & * \end{pmatrix} \right\}, \quad H' = \left\{ \begin{pmatrix} * & * \\ 0 & 1 \end{pmatrix} \right\}$$

of B. A simple calculation shows that

$$N_G(H) = N_G(H') = B.$$

We also have that for any $g \in G$,

$$gHg^{-1} \cap B \subseteq H' \text{ or } H.$$

We apply Theorem 9.3 to get a prime p which is unramified in L, the fixed field of the kernel of ρ_ℓ and which has the property that $\sigma_p \cap H = \phi$ and

$$p \ll x^2$$

where

$$x = \left\{ \frac{1}{\ell - 1 - 2} \frac{2}{\ell(\ell-1)} (\log |d_L|), \sqrt{\frac{\ell-1}{\ell^4}} (\log |d_L|) \right\}.$$

Now by Hensel's inequality,

$$\log|d_L| \ll n_L \log \ell \mathcal{N} \ll \ell^4 \log \ell \mathcal{N}$$

and so

$$p \ll (\ell \log \ell \mathcal{N})^2.$$

Now consider

$$D = \{g \in G : \mathrm{tr}\, g = 1 + \det g\}.$$

Clearly, every conjugacy class in D intersects H non-trivially. Hence σ_p is not contained in D, or in other words $\sigma_p \cap D = \phi$. Thus

$$a(p) \not\equiv 1 + p (\mathrm{mod}\, \ell)$$

and this means that

$$|E(\mathbb{F}_p)| = p + 1 - a(p) \not\equiv 0 (\mathrm{mod}\, \ell).$$

(ii): We may suppose (after a suitable choice of basis) that $G \subseteq B$ (with B as above).

We are again looking for a prime p such that $\sigma_p \cap H = \phi$ where

$$H = G \cap \left\{ \begin{pmatrix} 1 & * \\ 0 & * \end{pmatrix} \right\}.$$

If $G = H$, then it is clear that ℓ divides T and this is excluded by assumption. Thus, we may suppose that $G \neq H$.

Since H is a normal subgroup of G, it follows from Theorem 9.1 that there exists a prime p with the desired property and

$$p \ll \left(\frac{1}{[G:H]} \log d_F \right)^2$$

where F is the fixed field of H. Since F is a Galois extension of $\mathbb{Q}$ ramified only at primes dividing $\ell \mathcal{N}$, we have

$$p \ll (\log \ell \mathcal{N})^2.$$

(iii): This is impossible if $\ell > 2$ since G contains the image of complex conjugation, a matrix with distinct $\mathbb{F}_\ell$-rational eigenvalues (namely $+1, -1$), whereas the eigenvalues of every element of a nonsplit Cartan subgroup are either equal or lie in $\mathbb{F}_{\ell^2} \backslash \mathbb{F}_\ell$.

(iv): In this case $|G| \ll \ell$. By the result of Lagarias and Odlyzko , quoted as (9.1) at the beginning of this section, there exists a prime p whose σ_p is $\left\{ \begin{pmatrix} 2 & \\ & 2 \end{pmatrix} \right\}$ (say) and

$$p < (|G| \log \ell \mathcal{N})^2 \ll (\ell \log \ell \mathcal{N})^2.$$

Such a prime has

$$a(p) \equiv 4 \not\equiv 1 + 4 \equiv 1 + p \ (\mathrm{mod}\, \ell).$$

(v): In this case, there is a quadratic character ϵ with the property that

$$p \nmid \mathcal{N} \ \text{and} \ \epsilon(p) = -1 \Rightarrow a(p) \equiv 0 (\mathrm{mod}\, \ell).$$

Let K be the quadratic extension of $\mathbb{Q}$ corresponding to ϵ. This field has the property [Se2, p. 198] that it is unramified at ℓ and can only ramify at primes dividing $\mathcal{N}$. Hence, we can find a prime p such that $p \equiv 1 (\mathrm{mod}\, \ell)$ and $\epsilon(p) = -1$ with

$$p \ll (\log |d_{K(\zeta_\ell)}|)^2 \ll (\ell \log \ell \mathcal{N})^2$$

where ζ_ℓ is a primitive ℓ-th root of unity. For such a prime, $a(p) \equiv 0 \not\equiv 2 \equiv 1 + p (\mathrm{mod}\, \ell)$. This proves the theorem.

Exercises

1.   Let χ be an irreducible character of a finite group G. If χ is a linear combination with positive real coefficients of monomial characters, then $m\chi$ is monomial for some integer $m \geq 1$.

2.   Let A be a normal subgroup of the group G and let χ be an irreducible character of G. Then either the restriction of χ to A is isotypic (that is, a multiple of one character) or there is a subgroup H containing A and an irreducible character σ of H such that $\chi = \mathrm{Ind}_H^G \sigma$. (See [Se1, Prop. 24]).

3.   A finite group G is called *supersolvable* if there is a sequence of subgroups

$$\{1\} = G_0 \subseteq G_1 \subseteq \cdots G_n = G$$

with each G_i normal in G and with successive quotients G_i/G_{i-1} cyclic.
(a)   Prove that a nonabelian supersolvable group has a normal abelian subgroup which is not contained in the center.
(b)   Use (a) and Exercise 2 to prove that an irreducible character of a supersolvable group is monomial (that is, the induction of a one-dimensional character of some subgroup).

Exercises 4–7 are based on the paper [R] of Rhoades.

4.   Let $\mathcal{F}$ be a set of characters of the finite group G. We say that a class function
      θ is *semi-orthogonal* to $\mathcal{F}$ if $(\theta, \phi) \geq 0$ for all $\phi \in \mathcal{F}$.
      (a)   If $\mathcal{F}$ is the set of all characters of G, then a generalized character θ is
            semi-orthogonal to $\mathcal{F}$ if and only if θ is a character.
      (b)   Let

$$\tilde{\mathcal{F}} = \{\sum x_\phi \phi : 0 < x_\phi \in \mathbb{R}, \phi \in \mathcal{F}\}.$$

            Then a class function θ is semi-orthogonal to $\mathcal{F}$ if and only if it is semi-
            orthogonal to $\tilde{\mathcal{F}}$.

5.   If

$$\mathcal{F} = \{\mathrm{Ind}_H^G \psi : H \text{ an Abelian subgroup of } G\}$$

      then
      (a)   the generalized character θ_G is semi-orthogonal to $\mathcal{F}$.
      (b)   if a generalized character $\theta = \sum m_\chi \chi$ is semi-orthogonal to $\mathcal{F}$ then
            $|m_\chi| \leq |\theta(1)|$.

6.   Let F be a subset of $\mathbb{R}^k$ and define

$$\mathcal{H}(F) = \{x \in \mathbb{R}^k : (f, x) \geq 0 \text{ for all } f \in F\}$$

      and

$$\mathcal{C}(F) = \{\sum x_i f_i : 0 < x_i \in \mathbb{R}, f_i \in F\}$$

      where $(\ ,\ )$ denotes the standard inner product.
      (a)   If F is a subspace, then $\mathcal{H}(F)$ is the subspace of $\mathbb{R}^k$ orthogonal to F
            and $\mathcal{H}(\mathcal{H}(F)) = F$ and $\mathcal{C}(F) \subset F$.
     *(b)   [R, Lemma 1] If F does not contain the zero vector and all elements of
            F have non-negative coordinates, then $\mathcal{H}(\mathcal{H}(F)) = \mathcal{C}(F)$.

7.   Let G be a finite group and $\mathcal{F}$ a subset of characters of G. Expressing the
      elements of $\mathcal{F}$ as a sum of irreducible characters of G, identify $\mathcal{F}$ as a subset
      of $\mathbb{R}^k$ for some k. Using Exercise 6(b), show that a generalized character ψ of
      G can be written as a positive rational linear combination of characters in $\mathcal{F}$
      if and only if $(\psi, \theta) \geq 0$ for all θ semi-orthogonal to $\mathcal{F}$. Deduce that for any
      irreducible character χ of G, $\mathrm{reg}_G \pm \chi$ can be written as a positive rational
      linear combination of monomial characters.

8.   Let L/K be a finite Galois extension with group G. Show that the Artin
      L-functions $L(s, \chi, K)$ (as χ ranges over the irreducible characters of G) are
      multiplicatively independent over $\mathbb{Q}$. That is, if

$$\prod_\chi L(s, \chi, K)^{c_\chi} = 1$$

      for some rational numbers c_χ then $c_\chi = 0$ for all χ.

9. Let $F/\mathbb{Q}$ be a finite Galois extension with group G and let H, H' be two subgroups. Denote by K and K' the corresponding fixed fields.

   (a) Show that $\zeta_K(s) = \zeta_{K'}(s)$ if and only if for every conjugacy class C of G, we have $\#(H \cap C) = \#(H' \cap C)$.

   (b) Let $G = S_6$ (the symmetric group on 6 letters) and consider the subgroups

   $$H = \{(1), (12)(34), (12)(56), (34)(56)\}$$

   and

   $$H' = \{(1), (12)(34), (13)(24), (14)(23)\}.$$

   Prove that the above condition is satisfied and deduce that the Dedekind zeta functions of the corresponding fixed fields coincide. (This is due to Gassman, 1926.)

10. Let $1 < a \in \mathbb{Z}$ be a squarefree integer and q a prime. Set $K = \mathbb{Q}(a^{1/q})$ and prove directly that $\zeta_K(s)/\zeta(s)$ is entire.

11. Let $f(T) \in \mathbb{Z}[T]$ be an irreducible polynomial of degree larger than 1. Show that the set

    $$\{p : f(T) \equiv 0 (\bmod p) \text{ has a solution}\}$$

    has positive density.

12. Let E be a biquadratic extension of $\mathbb{Q}$ and let K_1, K_2, K_3 be the three quadratic subfields. Show that

    $$\zeta(s)^2 \zeta_E(s) = \zeta_{K_1}(s)\zeta_{K_2}(s)\zeta_{K_3}(s).$$

    Deduce a relation amongst the class numbers of the K_i.

References

[CF] J. Cassels and A. Fröhlich, Algebraic Number Theory, Academic Press, 1967.

[D] H. Davenport, Multiplicative Number Theory, Springer-Verlag, 1980.

[Fo] R. Foote, Non-monomial characters and Artin's conjecture, *Trans. Amer. Math. Soc.*, **321** (1990), 261–272.

[FM] R. Foote and V. Kumar Murty, Zeros and poles of Artin L-series, *Math. Proc. Camb. Phil. Soc.*, **105** (1989), 5–11.

[FW] R. Foote and D. Wales, Zeros of order 2 of Dedekind zeta functions and Artin's conjecture, *J. Algebra*, **131** (1990), 226–257.

[Fr] A. Fröhlich, Galois Module Structure of algebraic integers, Ergebnisse der Mathematik und ihrer Grenzgebiete, Springer-Verlag, 1983.

[K]   N. Katz, Galois properties of torsion points of abelian varieties, *Invent. Math.*, **62** (1981), 481–502.

[La]   S. Lang, Algebraic Number Theory, Springer-Verlag, 1986.

[LO]   J. Lagarias and A. M. Odlyzko, Effective versions of the Chebotarev Density Theorem, *Algebraic Number Fields*, ed. A. Fröhlich, 409–464, Academic Press, New York, 1977.

[LMO] J. Lagarias, H. Montgomery and A. M. Odlyzko, A bound for the least prime ideal in the Chebotarev Density Theorem, *Invent. Math.*, **54** (1979), 271–296.

[Mu1] V. Kumar Murty, Holomorphy of Artin L-functions, in: Proc. Ramanujan Centennial Conference, pp. 55–66, Ramanujan Mathematical Society, Chidambaram, 1987.

[Mu2] V. Kumar Murty, Explicit formulae and the Lang-Trotter conjecture, *Rocky Mountain J. Math.*, **15** (1985), 535–551.

[Mu3] V. Kumar Murty, The least prime which does not split completely, *Forum Math.*, **6** (1994), 555–565.

[MM]   M. Ram Murty and V. Kumar Murty, Base change and the Birch and Swinnerton-Dyer conjecture, in: A tribute to Emil Grosswald: number theory and analysis, *Contemp. Math.*, **143** (1993), 481–494.

[MMS] M. Ram Murty, V. Kumar Murty and N. Saradha, Modular forms and the Chebotarev Density Theorem, *Amer. J. Math.*, **110** (1988), 253–281.

[MS]   V. Kumar Murty and J. Scherk, Effective versions of the Chebotarev Density Theorem in the function field case, *C.R. Acad. Sci. Paris*, **319** (1994), 523–528.

[R]   S. Rhoades, A generalization of the Aramata-Brauer theorem, *Proc. Amer. Math. Soc.*, **119** (1993), 357–364.

[Se1] J.-P. Serre, Linear representations of finite groups, Springer-Verlag, New York, 1977.

[Se2] J.-P. Serre, Quelques applications du Théorème de Densité de Chebotarev, *Publ. Math. IHES*, **54** (1981), 123–201.

[St]   H. M. Stark, Some effective cases of the Brauer-Siegel theorem, *Invent. Math.*, **23** (1974), 135–152.

[Uc]   K. Uchida, On Artin L-functions, *Tohoku Math. J.*, **27** (1975), 75–81.

[vdW] R. W. van der Waall, On a conjecture of Dedekind on zeta functions, *Indag. Math.*, **37** (1975), 83–86.

Chapter 3
Equidistribution and L-Functions

§1 Compact groups and Haar measures

Let X be a compact topological space and $C(X)$ the Banach space of continuous, complex-valued functions on X, with the supremum norm:

$$\|f\| = \sup\big\{\, |f(x)| \mid x \in X \big\}.$$

Let x_1, x_2, x_3, ... be a sequence of points of X. Let μ be a Radon measure on X (that is, a continuous linear form on $C(X)$). The sequence x_1, x_2, x_3, ... is said to be μ-*equidistributed* if

$$\mu(f) = \lim_{n \to \infty} \frac{1}{n} \sum_{i=1}^{n} f(x_i).$$

We will now follow [Se] in our treatement.

Lemma 1.1   *Let ϕ_α be a family of continuous functions on X with the property that their linear combinations are dense in $C(X)$. Suppose that, for each α, the sequence $\mu_n(\phi_\alpha)$, $1 \le n < \infty$ where*

$$\mu_n(\phi_\alpha) = \frac{1}{n} \sum_{i=1}^{n} \phi_\alpha(x_i),$$

has a limit. Then the sequence x_1, x_2, x_3, ... is μ-equidistributed for some unique measure μ satisfying

$$\text{for all } \alpha: \ \mu(\phi_\alpha) = \lim_{n \to \infty} \mu_n(\phi_\alpha)$$

Proof. If $f \in C(X)$, a familiar argument using equicontinuity shows that the sequence $\mu_1(f)$, $\mu_2(f)$, ... has a limit $\mu(f)$, which is continuous and linear in f. This proves the lemma.

Lemma 1.2 *Suppose that x_1, x_2, x_3, ... is μ-equidistributed. Let U be a subset of X whose boundary has μ-measure zero and for all n, let n_U be the number of $m \leq n$ such that $x_m \in U$. Then*

$$\lim_{n \to \infty} \frac{n_U}{n} = \mu(U).$$

Proof. We normalize our measure so that $\mu(X) = 1$. Let U^0 be the interior of U. We have $\mu(U^0) = \mu(U)$. Let $\epsilon > 0$. By the definition of $\mu(U^0)$, there is a continuous function $\phi \in C(X)$, $0 \leq \phi \leq 1$, with $\phi = 0$ on $X - U^0$ and $\mu(\phi) \geq \mu(U) - \epsilon$. Since $\mu_n(\phi) \leq n_U/n$ we have

$$\lim_{n \to \infty} \inf n_U/n \geq \lim_{n \to \infty} \mu_n(\phi) = \mu(\phi) \geq \mu(U) - \epsilon,$$

from which we obtain $\lim \inf n_U/n \geq \mu(U)$. The same argument applied to $X - U$ shows that

$$\lim_{n \to \infty} \inf (n - n_U)/n \geq \mu(X - U).$$

Hence,

$$\lim_{n \to \infty} \sup n_U/n \leq \mu(U) \leq \lim_{n \to \infty} \inf n_U/n.$$

which implies the lemma.

Example. If $X = [0, 1]$ and μ is the usual Lebesgue measure, a sequence $\{x_n\}_{n=1}^{\infty}$ of points of X is μ-equidistributed if and only if for each interval $[a, b]$ of length d in $[0, 1]$ the number of $m \leq n$ such that $x_m \in [a, b]$ is equal to $dn + \mathbf{o}(n)$ as $n \to \infty$.

§2 Weyl's criterion for equidistribution

If G is a compact group, let X denote the space of conjugacy classes of G (that is, the quotient space of G by the equivalence relation induced by inner automorphisms of G). Let μ be a measure on G; its image under the quotient map $G \to X$ is a measure on X which we also denote by μ.

Proposition 2.1 *Let G be a compact group, X its space of conjugacy classes. Let μ be a measure on G. The sequence $\{x_n\}_{n=1}^{\infty}$ is μ-equidistributed if and only if for any irreducible character χ of G, we have,*

$$\lim_{n \to \infty} \frac{1}{n} \sum_{i=1}^{n} \chi(x_i) = \mu(\chi).$$

Proof. The map $C(X) \to C(G)$ is an isomorphism of $C(X)$ onto the space of class functions on G; by the Peter-Weyl Theorem, the irreducible characters χ of G generate a dense subspace of $C(X)$. Hence the proposition follows from Lemma 1.1.

Corollary 2.2 *Let μ be the Haar measure of G with $\mu(G) = 1$. Then the sequence $\{x_n\}_{n=1}^{\infty}$ of elements of X is μ-equidistributed if and only if for any irreducible character χ of G, $\chi \neq 1$, we have,*

$$\lim_{n \to \infty} \frac{1}{n} \sum_{i=1}^{n} \chi(x_i) = 0.$$

Proof. This follows from Proposition 2.1 and (a special case of) the orthogonality relations: $\mu(\chi) = 0$ if χ is irreducible $\neq 1$ and $\mu(1) = 1$.

Corollary 2.3 (H. Weyl) *Let $G = \mathbb{R}/\mathbb{Z}$, and let μ be the normalized Haar measure on G. Then $\{x_n\}_{n=1}^{\infty}$ is μ-equidistributed if and only if for any integer $m \neq 0$ we have*

$$\sum_{n \leq N} e^{2\pi i m x_n} = \circ(N)$$

as $N \to \infty$.

Proof. It suffices to remark that the irreducible characters of $\mathbb{R}/\mathbb{Z}$ are the maps $x \mapsto e^{2\pi i m x}$, $(m \in \mathbb{Z})$.

In [BTD], the image of the Haar measure of $SU(2, \mathbb{C})$ in the space of conjugacy classes is calculated. Each conjugacy class has a representative of the form

$$\begin{pmatrix} e^{i\theta} & 0 \\ 0 & e^{-i\theta} \end{pmatrix}, 0 \leq \theta \leq \pi$$

which has measure $\frac{2}{\pi} \sin^2 \theta d\theta$.

§3 *L*-functions on *G*

Let G be a compact group, X its space of conjugacy classes as above. Suppose that for each prime p, we associate a conjugacy class X_p in X. As p varies, how are the X_p distributed (say, with respect to the normalised Haar measure on X)? To answer this, we define the *L*-function associated to each irreducible complex linear representation ρ of G in the following way: Let

$$\rho : G \longrightarrow GL_n(\mathbb{C})$$

and define

$$L(s, \rho) = \prod_p \det(1 - \rho(X_p)p^{-s})^{-1}.$$

Theorem 3.1 *Suppose that for each irreducible representation $\rho \neq 1$, of G, the L-function $L(s, \rho)$ extends to an analytic function for $\mathrm{Re}(s) \geq 1$, and does not vanish there. Then the X_p's are uniformly distributed with respect to the image of the Haar measure in the space of the conjugacy classes of G.*

Proof. By the Wiener-Ikehara Tauberian Theorem 1.1 in Chapter 1, together with corollary 2.2 above, the result follows immediately.

Example.   Let ζ_n be a primitive n-th root of unity and set $K = \mathbb{Q}(\zeta_n)$, be the usual n-th cyclotomic field. Then

$$Gal(K/\mathbb{Q}) \simeq (\mathbb{Z}/n\mathbb{Z})^*.$$

Moreover, for each prime $p \nmid n$, σ_p is defined and

$$\sigma_p(\zeta_n) = \zeta_n^p.$$

Thus σ_p depends only on the arithmetical progression to which p belongs mod n. Hence, if $\pi(x; n, a)$ is the number of primes $p \leq x$, $p \equiv a \pmod{n}$ then

$$\pi(x; n, a) \sim \frac{1}{\phi(n)} \ell i \, x$$

as $x \to \infty$.

The L-functions attached to G in this example are the classical Dirichlet L-functions. We therefore obtain the prime number theorem for arithmetic progressions if for $t \in \mathbb{R}$,

$$L(1 + it, \chi) \neq 0.$$

Hence, the prime number theorem for arithmetic progressions, and more generally, the Chebotarev density theorem, fit into this general formalism. In [Se2], Serre formulates a 'motivic' generalization of the Chebotarev density theorem.

§4 Deligne's Prime Number Theorem

Let G be a compact group. An irreducible character χ of G will be called **quadratic** if its degree is 1 and its image consists of ± 1.

Theorem 4.1 *Let G be a compact group. Assume that for every non-trivial, irreducible representation ρ of G, $L(s, \rho)$ is holomorphic at $s = 1$. If χ is quadratic we suppose that $L(s, \chi)$ is holomorphic on $[1/2, 1]$. Then $L(1, \rho) \neq 0$ for all irreducible $\rho \neq 1$.*

Remark. We want to point out that the second condition above is essential. For instance, consider the group $G = \pm 1$ of order 2. For each prime p, we will define $X_p = -1$. G has only two irreducible characters, the trivial one which gives the Riemann zeta function and the non-trivial one which gives

$$\prod_p \frac{1}{1+p^{-s}} = \prod_p \frac{1-p^{-s}}{1-p^{-2s}} = \frac{\zeta(2s)}{\zeta(s)}.$$

This function is holomorphic at $s = 1$ but not at $s = 1/2$. However, it does vanish at $s = 1$.

Before we prove the theorem, we establish the following which is a variation of the theme of Hadamard and de la Vallée Poussin.

Lemma 4.2 *Let χ be an irreducible character of G. Then there exists a function f on G of the form*

$$f = \sum_\psi c_\psi \psi$$

which is a sum over a finite set of irreducible characters ψ where $c_\psi \in \mathbb{Z}$. Moreover, $\mathrm{Re}(f) \geq 0$, $f \neq 0$ and $c_1 \leq c_\chi$. If $\chi \neq 1$ is not quadratic, then $c_1 < c_\chi$.

Proof. We consider three cases. If $\chi = 1$, then let $f = 1$. If χ is quadratic then let $f = 1 + \chi$. If χ is a character of degree 1, which is not quadratic, then let $f = 3 + 4\chi + \chi^2$ and we see that $\mathrm{Re}(f) \geq 0$ by the inequality of Hadamard and de la Vallée Poussin. Now let χ be a character of degree $d > 1$. If G is finite, we could take

$$f = \sum_\psi \psi(1)\psi$$

which is the character of the regular representation. We find that

$$f(g) = \begin{cases} |G| & \text{if } g = 1 \\ 0 & \text{otherwise.} \end{cases}$$

In the case of an arbitrary compact group, we will try and approximate this construction.

Given any $\epsilon > 0$, there exists a delta like function F on G such that F is continuous, real-valued, non-negative, invariant under conjugation, and $F(x) = F(x^{-1})$ satisfying

$$\int_G F(x)dx = 1,$$

and

$$\int_G F(x)\chi(x)dx > d - \epsilon$$

for $\epsilon > 0$. Note that the latter quantity is a real number since it is invariant under complex conjugation. Now choose a finite sum that approximates F:

$$\sum_\psi c_\psi \psi.$$

We may assume without loss that c_ψ are rational numbers. Thus we have a function

$$F = \sum_\psi c_\psi \psi$$

which is a finite sum, with c_ψ rational numbers, real valued and non-negative, $c_1 = 1$ and $c_\chi > d - \epsilon$. Now take $f = F * F$ to get non-negative coefficients so that

$$f(x) = \int_G F(xy^{-1})F(y)\,dy.$$

Moreover, if ψ is irreducible, the orthogonality relations yield

$$\psi * \psi = \frac{1}{\psi(1)}\psi$$

so that

$$f = \sum \frac{c_\psi^2}{\psi(1)}\psi.$$

The coefficients are still rational numbers. The coefficient of the trivial character is 1 and of χ is $> (d-\epsilon)^2/d$. Clearing denominators gives us the desired function.

Proof of Theorem 4.1 Assume that ρ is not one or quadratic. Choose $f = \sum_\psi c_\psi \psi$ as in the lemma with $c_1 < c_\chi$. Assume $L(1,\chi) = 0$. Then,

$$L(s,f) = \prod_\psi L(s,\psi)^{c_\psi}$$

$$= L(s,1)^{c_1} L(s,\chi)^{c_\chi} \prod_{\psi \neq 1,\chi} L(s,\psi)^{c_\psi},$$

so that $L(s,f)$ has a zero at $s = 1$. Therefore, for some positive integer m,

$$\frac{-L'}{L}(s,f) = \frac{-m}{s-1} + \mathbf{O}(1)$$

as $s \to 1^+$. Hence,

$$\lim_{s \to 1^+} -\operatorname{Re}(\frac{L'}{L}(s,f)) < 0.$$

On the other hand, $L(s, f)$ has non-negative coefficients and therefore,

$$- \text{Re}(\frac{L'}{L}(s, f)) \geq 0$$

which implies

$$\lim_{s \to 1^+} - \text{Re}(\frac{L'}{L}(s, f)) \geq 0$$

which is a contradiction. This completes the proof in this case. If χ is quadratic, consider

$$f(s) = L(s, \chi)L(s/2, 1)L(s/2, \chi)$$

and suppose that $L(1, \chi) = 0$. By hypothesis, $f(s)$ is holomorphic for s in $[1, 2]$ with a zero at $s = 1$. It is easily verified by looking at the Euler factors that the coefficients of $-f'/f$ are non-negative. Call β the zero of f on $[1, 2]$ closest to 2. Such a zero exists since $f(1) = 0$. Therefore, for some positive integer m,

$$-\frac{f'}{f}(s) = -\frac{m}{s - \beta} + \mathbf{O}(1)$$

as $s \to \beta$. But Landau's lemma (Exercise 5 in Chapter 1) implies

$$-\frac{f'}{f} \to \infty$$

as $s \to \beta$. This is a contradiction. Therefore $L(1, \chi) \neq 0$ and this completes the proof.

Exercises

1.   Prove that the sequence $\{\log p\}$ as p varies over the prime numbers is not uniformly distributed mod 1.

2.   (Erdös-Turán inequality): Let $\{x_j\}_{j=1}^n$ be a finite sequence of real numbers. For $0 \leq \alpha < \beta \leq 1$. Let $A([\alpha, \beta] : n)$ be the counting function

$$\#\{j \leq n : x_j \in [\alpha, \beta](\text{mod } 1)\}.$$

Define the discrepancy

$$D_n(x_1, \ldots, x_n) = \sup_{0 \leq \alpha < \beta \leq 1} \left| \frac{A([\alpha, \beta] : n)}{n} - (\beta - \alpha) \right|.$$

Then, for any positive integer M,

$$D_n \leq \frac{6}{M + 1} + \frac{4}{\pi} \sum_{h=1}^{M} \left(\frac{1}{h} - \frac{1}{M + 1} \right) \left| \frac{1}{n} \sum_{j=1}^{n} e^{2\pi i h x_j} \right|.$$

Recently, Montgomery [Mo] obtained the following improvement:

$$D_n \le \frac{1}{M+1} + 2\sum_{h=1}^{M}\left(\frac{1}{M+1} + \min(\beta - \alpha, \frac{1}{\pi h})\right)\frac{1}{n}\sum_{j=1}^{n}e^{2\pi i h x_j}.$$

*3. Let $\mathbb{F}_p$ denote the finite field of p elements. Let $\psi : \mathbb{F}_p \to \mathbb{C}^*$ be a fixed non-trivial additive character. For each character χ of the multiplicative group $\mathbb{F}_p^*$, we define a Gauss sum

$$G(\chi, \psi) = \sum_{x \in \mathbb{F}_p^*} \chi(x)\psi(x).$$

If $\chi \ne 1$, we know that $|G(\chi, \psi)| = p^{1/2}$ so we can write

$$G(\chi, \psi) = p^{1/2}e^{2\pi i \theta_p(\chi)}.$$

Show that the sequence obtained by listing $\theta_p(\chi)$ as p ranges over all the prime numbers and for each fixed p, χ ranges over all the non-trivial characters mod p, is uniformly distributed mod 1.

4. Suppose that

$$c_\chi = \lim_{n\to\infty}\frac{1}{n}\sum_{i=1}^{n}\chi(x_i).$$

Define

$$\Phi(g) = \sum_\chi c_\chi \chi(g),$$

where the sum is over all irreducible characters arranged by increasing degrees. Then determine conditions when $\{x_n\}_{n=1}^{\infty}$ is uniformly distributed with respect to the measure $\Phi(g)d\mu$ where μ is the normalized Haar measure of G.

5. In Theorem 4.1, let us suppose further that each $L(s, \rho)$ is holomorphic on $\mathrm{Re}(s) = 1$ with $s \ne 1$. Show that $L(1 + it, \rho) \ne 0$ for $t \in \mathbb{R}$, $t \ne 0$. (Hint: for $t \in \mathbb{R}$ and non-zero, consider $S_t = \mathbb{R}/t\mathbb{Z}$ and the group $G \times S_t$. Define for each prime p the conjugacy class $(X_p, \log p)$ and now apply Theorem 4.1 to the L-functions associated to $G \times S_t$.)

6. For each prime $p \equiv 1 \pmod 4$, let $\chi \bmod p$ be a character of order 4 and define

$$J(\chi, \chi) = \sum_{\substack{x \bmod p \\ x \ne 0,1}} \chi(x)\chi(1 - x).$$

Show that if we write $J(\chi, \chi) = a + bi$ with $a, b \in \mathbb{Z}$, then $a^2 + b^2 = p$. Writing

$$J(\chi, \chi) = \sqrt{p}e^{i\theta_p}$$

show that the sequence of θ_p's is uniformly distributed mod 1.

7. Show that the set of primes p which can be written as

$$x^2 + xy + 6y^2$$

with $x, y \in \mathbb{Z}$ has density $1/6$. (Hint: consider $\mathbb{Q}(\sqrt{-23})$.)

*8. In the above question, define ϕ_p by

$$\tan \phi_p = \frac{y\sqrt{23}}{2x + y}.$$

Show that the ϕ_p's are equidistributed mod π.

9. Show that the density of primes represented by $x^2 - 2y^2$ is $1/2$. (Hint: consider $\mathbb{Q}(\sqrt{2})$).

*10. For each prime in question 9, show that the quantity

$$X_p = \log(|x| + |y|\sqrt{2})$$

is well defined modulo $\log(3 + 2\sqrt{2})$. Show that the sequence is uniformly distributed mod $\log(3 + 2\sqrt{2})$.

References

[BTD] T. Bröcker, T. tom Dieck, Representations of compact Lie groups, Graduate texts in mathematics, 98, Springer-Verlag, 1985.

[Mo] H. Montgomery, Ten lectures on the interface between analytic number theory and harmonic analysis, No. 84, CBMS, Amer. Math. Soc. 1994.

[Se] J.-P. Serre, *Abelian ℓ-adic representations and elliptic curves*, Benjamin, New York, 1968; second edition, Addison-Wesley, Redwood City, 1989.

[Se2] J.-P. Serre, Propriétés conjecturales des groupes de Galois motiviques et des représentations ℓ-adiques in Proc. Symp. Pure Math., **55** (1994) p. 377–400.

Chapter 4
Modular Forms and Dirichlet Series

§1 $SL_2(\mathbb{Z})$ and some of its subgroups

It was Ramanujan who in a fundamental paper of 1916 introduced his τ-function as the Fourier coefficient of a modular form and then attached a Dirichlet series to it. He established an analytic continuation of the series and a functional equation for it. He then made his famous conjectures about the multiplicativity of these coefficients and their size. The multiplicativity conjecture would allow him to write his Dirichlet series as an Euler product thereby establishing an analogy with classical zeta and L-functions. Subsequently Mordell proved that $\tau(n)$ is a multiplicative function but it was left to Hecke to develop a more elaborate theory and establish the existence of an infinite family of such examples. Ramanujan's conjecture on estimating the size of $\tau(n)$ however defied immediate attack. The fundamental method of Rankin and Selberg did allow one to get good estimates for them but they were not optimal. The final resolution of Ramanujan's conjecture came from algebraic geometry when it was shown to be a consequence of Deligne's proof of the celebrated Weil conjectures. In this chapter, we will give a brief introduction to the fundamental concepts and study the oscillations of the Fourier coefficients from the standpoint of the non-vanishing of various L-functions.

We begin with some basic notions. If R is any commutative ring with identity, $SL_2(R)$ denotes the group of matrices

$$\begin{pmatrix} a & b \\ c & d \end{pmatrix}$$

of determinant 1 and $a, b, c, d \in R$. We will consider the case $R = \mathbb{Z}$ and look at some subgroups of $SL_2(\mathbb{Z})$.

The principal congruence subgroup of level N is denoted $\Gamma(N)$ and consists of matrices in $SL_2(\mathbb{Z})$ satisfying

$$\begin{pmatrix} a & b \\ c & d \end{pmatrix} \equiv \begin{pmatrix} 1 & 0 \\ 0 & 1 \end{pmatrix} \pmod{N}$$

Since this is the kernel of the natural map (reduction $\mathrm{mod}\, N$):

$$SL_2(\mathbb{Z}) \quad \rightarrow \quad SL_2\left(\mathbb{Z}/N\mathbb{Z}\right)$$

$\Gamma(N)$ is a normal subgroup of finite index in $SL_2(\mathbb{Z})$.

The *Hecke subgroup* of level N is denoted $\Gamma_0(N)$ and consists of matrices

$$\begin{pmatrix} a & b \\ c & d \end{pmatrix} \quad \in \quad SL_2(\mathbb{Z})$$

such that $N \mid c$. Since

$$\Gamma(N) \subseteq \Gamma_0(N) \subseteq SL_2(\mathbb{Z}) \ ,$$

$\Gamma_0(N)$ has finite index in $SL_2(\mathbb{Z})$.

The group $\Gamma_1(N)$ consists of matrices $\gamma \in SL_2(\mathbb{Z})$ satisfying

$$\gamma \equiv \begin{pmatrix} 1 & * \\ 0 & 1 \end{pmatrix} \quad (\mathrm{mod}\, N)$$

Clearly

$$\Gamma(N) \subseteq \Gamma_1(N) \subseteq \Gamma_0(N) \subseteq SL_2(\mathbb{Z})$$

A *congruence subgroup* of $SL_2(\mathbb{Z})$, is by definition, a subgroup which contains $\Gamma(N)$ for some N. $\Gamma_0(N), \Gamma_1(N)$ are examples of congruence subgroups.

An element $\gamma \in SL_2(\mathbb{Z})$ is called *elliptic* if $|\mathrm{tr}\ \gamma| < 2$, *parabolic* if $|\mathrm{tr}\ \gamma| = 2$ and *hyperbolic* if $|\mathrm{tr}\ \gamma| > 2$.

§2 The upper half-plane

Let $\mathfrak{h}$ denote the upper half-plane:

$$\mathfrak{h} = \left\{ z = x + \sqrt{-1}y : x \in \mathbb{R}, y > 0 \right\}$$

Let $GL_2^+(\mathbb{R})$ be the group of 2×2 matrices with real entries and positive determinant. Then $GL_2^+(\mathbb{R})$ acts on $\mathfrak{h}$ as a group of holomorphic automorphisms:

$$\gamma : z \mapsto \frac{az+b}{cz+d}, \quad \gamma = \begin{pmatrix} a & b \\ c & d \end{pmatrix} \in GL_2^+(\mathbb{R})$$

Let $\mathfrak{h}^*$ denote the union of $\mathfrak{h}$ and the rational numbers $\mathbb{Q}$ together with a symbol ∞ (or more suggestively $i\infty$). The action of $SL_2(\mathbb{Z})$ on $\mathfrak{h}$ can be extended to $\mathfrak{h}^*$ by defining

$$\begin{pmatrix} a & b \\ c & d \end{pmatrix} \cdot \infty = \frac{a}{c} \quad \text{for} \ \ c \neq 0$$

and

$$\begin{pmatrix} a & b \\ 0 & d \end{pmatrix} \cdot \infty = \infty;$$

for rational numbers $\frac{r}{s}$ with $(r, s) = 1$, we define

$$\begin{pmatrix} a & b \\ c & d \end{pmatrix} \cdot \frac{r}{s} = \frac{ar + bs}{cr + ds}$$

with the understanding that when $cr + ds = 0$, the right side of the above equation is the symbol ∞. The rational numbers together with ∞ are called *cusps*.

If Γ is a discrete subgroup of $SL_2(\mathbb{R})$, then the orbit space $\mathfrak{h}^*/\Gamma$ can be given the structure of a compact Riemann surface, and is denoted X_Γ.

We will be interested in the case Γ is a congruence subgroup of $SL_2(\mathbb{Z})$. In that case, the algebraic curve corresponding to X_Γ is called a *modular curve*.

In case $\Gamma = \Gamma(N), \Gamma_1(N)$ or $\Gamma_0(N)$, the corresponding modular curve is denoted $X(N), X_1(N)$ and $X_0(N)$ respectively. For further details, the reader may consult [VKM].

§3 Modular forms and cusp forms

Let f be a holomorphic function on $\mathfrak{h}$ and k a positive integer. For

$$\gamma = \begin{pmatrix} a & b \\ c & d \end{pmatrix} \in GL_2^+(\mathbb{R}),$$

define

$$(f|_k\gamma)(z) = (\det\gamma)^{k/2}(cz + d)^{-k} f\left(\frac{az + b}{cz + d}\right).$$

For fixed k, the map $\gamma : f \mapsto f|_k\gamma$ defines an action of $GL_2^+(\mathbb{R})$ on the space of holomorphic functions on $\mathfrak{h}$. (Sometimes, we simply write $f|\gamma$ for $f|_k\gamma$.)

Let Γ be a subgroup of finite index in $SL_2(\mathbb{Z})$. Let f be a holomorphic function on $\mathfrak{h}$ such that $f|_k\gamma = f$ for all $\gamma \in \Gamma$. Since Γ has finite index,

$$\begin{pmatrix} 1 & 1 \\ 0 & 1 \end{pmatrix}^M = \begin{pmatrix} 1 & M \\ 0 & 1 \end{pmatrix} \in \Gamma$$

for some positive integer M. Hence $f(z + M) = f(z)$ for all $z \in \mathfrak{h}$. So, f has a "Fourier expansion at infinity":

$$f(z) = \sum_{n=-\infty}^{\infty} a_n q_M^n, \qquad q_M = e^{2\pi iz/M}$$

We say that f *is holomorphic at infinity* if $a_n = 0$ for all $n < 0$. We say it *vanishes at infinity* if $a_n = 0$ for all $n \le 0$.

Let $\sigma \in SL_2(\mathbb{Z})$. Then $\sigma^{-1}\Gamma\sigma$ also has finite index and $(f|\sigma)|\gamma = f|\sigma$ for all $\gamma \in \sigma^{-1}\Gamma\sigma$. So for any $\sigma \in SL_2(\mathbb{Z})$, $f|\sigma$ also has a Fourier expansion at infinity. We say that f is *holomorphic at the cusps* if $f|\sigma$ is holomorphic at infinity for all $\sigma \in SL_2(\mathbb{Z})$.

We say that f *vanishes at the cusps* if $f|\sigma$ vanishes at infinity for all $\sigma \in SL_2(\mathbb{Z})$.

Let N be an integer ≥ 1 and ϵ a Dirichlet character mod N. A *modular form on $\Gamma_0(N)$ of type (k, ϵ)* is a holomorphic function f on $\mathfrak{h}$ such that

(i)
$$f\Big|\begin{pmatrix} a & b \\ c & d \end{pmatrix} = \epsilon(d)f \text{ for all } \begin{pmatrix} a & b \\ c & d \end{pmatrix} \in \Gamma_0(N)$$

and

(ii) f is holomorphic at the cusps.

(Note that (i) implies $f|\gamma = f$ for all $\gamma \in \Gamma_1(N)$ and so (ii) is meaningful.)

Also the Fourier expansion of such a form is :

$$f(z) = \sum_{n=0}^{\infty} a_n\, q^n, \qquad\qquad q = e^{2\pi i z}$$

The integer k is called the *weight* of f. Such a modular form is called a *cusp form* if it vanishes at the cusps.

The modular forms on $\Gamma_0(N)$ of type (k, ϵ) form a complex vector space $M_k\left(\Gamma_0(N), \epsilon\right)$ and this has a subspace $S_k\left(\Gamma_0(N), \epsilon\right)$ consisting of cusp forms. The subspace has a canonical complement:

$$M_k\left(\Gamma_0(N), \epsilon\right) = \mathcal{E}_k\left(\Gamma_0(N), \epsilon\right) \bigoplus S_k\left(\Gamma_0(N), \epsilon\right)$$

and the space $\mathcal{E}_k$ is called the space spanned by Eisenstein series. These spaces are finite dimensional. Moreover, one can define an *inner product* (Petersson) on $S_k\left(\Gamma_0(N), \epsilon\right)$ by

$$\langle f, g \rangle = \int_{\mathfrak{h}/\Gamma_0(N)} f(z)\overline{g(z)}y^k \frac{dxdy}{y^2}$$

Examples

1. Let $k \ge 4$ be even. Then

$$G_k(z) = \sum_{\substack{m,n \in \mathbb{Z} \\ (m,n) \ne (0,0)}} (mz + n)^{-k}$$

is a modular form of weight k for $SL_2(\mathbb{Z})$. Its Fourier expansion is

$$G_k(z) = 2\zeta(k) + 2\frac{(2\pi i)^k}{(k-1)!}\sum_{n=1}^{\infty}\sigma_{k-1}(n)\,q^n$$

where $\sigma_k(n) = \sum_{d|n}d^k$, and ζ denotes the Riemann zeta function. If we normalize G_k so that the constant term is 1, and use the well-known formula

$$2\zeta(k) = -\frac{B_k}{k!}(2\pi i)^k$$

we find

$$E_k(z) = 1 - \frac{2k}{B_k}\sum_{n=1}^{\infty}\sigma_{k-1}(n)q^n$$

Here B_k is the k-th Bernoulli number defined by

$$\frac{t}{e^t-1} = \sum_{k=0}^{\infty}\frac{B_k t^k}{k!}$$

E_k is called the k-th Eisenstein series.

2. Ramanujan's cusp form Define

$$\Delta(z) = q\prod_{n=1}^{\infty}(1-q^n)^{24}, q = e^{2\pi iz}$$

Then

$$\Delta(z) = \sum_{n=1}^{\infty}\tau(n)\,q^n$$

is a cusp form of weight 12 for $SL_2(\mathbb{Z})$; $\tau(n)$ is called the Ramanujan function. Ramanujan conjectured in 1916 that

(i) $\tau(nm) = \tau(n)\tau(m)$ $\quad (n,m) = 1$
(ii) $|\tau(p)| \le 2p^{11/2},$ $\quad p$ prime

(i) was proved by Mordell in 1928 and (ii) by Deligne in 1974. More generally, if $f \in S_k(\Gamma_0(N), \epsilon)$, then its Fourier coefficients satisfy

$$a_n = O(n^{\frac{k-1}{2}+\delta}) \quad \text{for any } \delta > 0.$$

For $k \ge 2$, this is due to Deligne. If $k = 1$, this is a theorem of Deligne-Serre.

3. Define $B_{n,\chi}$ by

$$\sum_{a=1}^{c}\chi(a)\frac{te^{at}}{e^{ct}-1} = \sum_{n=0}^{\infty}\frac{B_{n,\chi}t^n}{n!}$$

where χ is a Dirichlet character modulo c. Let

$$E_{1,\chi} = 1 - \frac{2}{B_{1,\chi}} \sum_{n=1}^{\infty} \left(\sum_{d|n} \chi(d) \right) q^n$$

If χ is *odd* (i.e., $\chi(-1) = -1$), then $E_{1,\chi} \in M_1 (\Gamma_0(c), \chi)$. One can also show that if χ is *even* (i.e., $\chi(-1) = 1$) then

$$E_{2,\chi} = 1 - \frac{4}{B_{2,\chi}} \sum_{n=1}^{\infty} \left(\sum_{d|n} \chi(d)d \right) q^n$$

is in $M_2 (\Gamma_0(c), \chi)$.

For higher weights, one has analogous results. See for instance [La].

Theorem 3.1   *Let $S_k(N) = S_k(\Gamma_0(N), 1)$. Then*

$$\dim S_2(N) = 1 + \frac{i(N)}{12} - \frac{i_2(N)}{4} - \frac{i_3(N)}{3} - \frac{i_\infty(N)}{2}$$

where

$$i(N) = N \prod_{p|N} \left(1 + \frac{1}{p} \right)$$

$$i_2(N) = \begin{cases} 0 & \text{if } 4 \mid N \\ \prod_{p|N} \left(1 + \left(\frac{-4}{p} \right) \right) & \text{otherwise} \end{cases}$$

$$i_3(N) = \begin{cases} 0 & \text{if } 2 \mid N \text{ or } 9 \mid N \\ \prod_{p|N} \left(1 + \left(\frac{-3}{p} \right) \right) & \text{otherwise} \end{cases}$$

$$i_\infty(N) = \sum_{d|N} \phi\left((d, N/d)\right)$$

This formula occurs in [Kn, p. 272] with a misprint which has been corrected above. One can write down analogous formulas for $\dim S_k(N)$ for $k \geq 2$. (See for example, [Shi] or [CO].) For explicit examples, see Frey [F]. For $k = 1$, no such simple formula is known. However, for N prime, it is conjectured [D] that

$$\dim S_1(N) = \frac{1}{2}h(-N) + \mathbf{O}(N^\epsilon)$$

where $h(-N)$ denotes the class number of $\mathbb{Q}(\sqrt{-N})$.

§4 *L*-functions and Hecke's theorem

If $f \in S_k\left(\Gamma_0(N), \epsilon\right)$ and $f(z) = \sum_{n=1}^{\infty} a_n e^{2\pi i n z}$ is its Fourier expansion at $i\infty$, we attach an *L*-function by

$$L(s, f) = \sum_{n=1}^{\infty} \frac{a_n}{n^s}$$

Since $y^{k/2}|f(z)|$ is invariant under $\Gamma_1(N)$, and hence represents a function on the compact Riemann surface $X_1(N)$, it is bounded on $\mathfrak{h}$. Therefore

$$e^{-2\pi n y} a_n = \int_{-1/2}^{1/2} f(x + iy) e^{-2\pi i n x} dx$$
$$\ll y^{-k/2}$$

Setting $y = 1/n$ gives $a_n = O(n^{k/2})$.

This shows that $L(s, f)$ represents an analytic function for $\mathrm{Re}\, s > \frac{k+2}{2}$.

Let $W_N = \begin{pmatrix} 0 & -1 \\ N & 0 \end{pmatrix}$. It is *not* an element of $SL_2(\mathbb{Z})$. However,

$$W_N \Gamma_0(N) W_N^{-1} \subset \Gamma_0(N)$$

and so $f \mapsto f|W_N$ preserves $M_k\left(\Gamma_0(N)\right)$ and $S_k\left(\Gamma_0(N)\right)$. Moreover, $f|W_N^2 = f$. W_N is called the *Atkin-Lehner involution*.

Since W_N is a linear transformation of the vector space $S_k\left(\Gamma_0(N)\right)$ and $W_N^2 = 1$, it decomposes the space into $S_k^+\left(\Gamma_0(N)\right)$ and $S_k^-\left(\Gamma_0(N)\right)$ corresponding to the eigenvalues ± 1. Note that if $S_k(\Gamma_0(N)) \neq 0$ then k is even.

Theorem 4.1   (Hecke) *Let $f \in S_k^{\pm}\left(\Gamma_0(N)\right)$. Then $L(s, f)$ extends to an entire function and*

$$\Lambda(s, f) = N^{s/2}(2\pi)^{-s}\Gamma(s)L(s, f)$$

satisfies the functional equation

$$\Lambda(s, f) = \pm(-1)^{k/2}\Lambda(k - s, f)$$

Proof. Since $f|W_N = \pm f$, we find

$$f\left(\frac{i}{Ny}\right) = \pm N^{k/2} i^k y^k f(iy)$$

Since

$$N^{-s/2}\Lambda(s, f) = \int_0^{\infty} f(iy) y^{s-1} dy$$

we see that

$$N^{-s/2}\Lambda(s, f) = \int_0^{1/\sqrt{N}} f(iy) y^{s-1} dy + \int_{1/\sqrt{N}}^{\infty} f(iy) y^{s-1} dy$$

In the first term, we replace y by $1/Ny$ and use the modular relation to get

$$N^{-s/2}\Lambda(s,f) = \pm N^{k/2}i^k \int_{1/\sqrt{N}}^\infty f(iy)y^{k-s-1}dy + \int_{1/\sqrt{N}}^\infty f(iy)y^{s-1}dy$$

which gives the analytic continuation and functional equation.

Corollary 4.2   Let $f \in S_k(\Gamma_0(N))$. Then $L(s,f)$ extends to an entire function.

§5 Hecke operators

Let p denote a prime number and $f(z) = \sum_{n=0}^\infty a_n q^n$ be a modular form on $\Gamma_0(N)$ of type (k,ϵ). The *Hecke operators* T_p and U_p are defined by

$$f|T_p = \sum_{n=0}^\infty a_{np}q^n + \epsilon(p)p^{k-1}\sum_{n=0}^\infty a_n q^{np} \qquad\qquad \text{if } p \nmid N$$

$$f|U_p = \sum_{n=0}^\infty a_{np}q^n \qquad\qquad \text{if } p \mid N$$

It is not difficult to show that $f|T_p$, $f|U_p$ are also modular forms on $\Gamma_0(N)$ of type (k,ϵ), and they are cusp forms if f is a cusp form.

Theorem 5.1   (Hecke)   *The T_p's are commuting linear transformations of $S_k(\Gamma_0(N),\epsilon)$. As such, the space can be decomposed as a direct sum of eigenspaces.*

Let $f \in S_k(\Gamma_0(N),\epsilon)$. We will say that f is an *eigenform* if f is an eigenfunction for all the Hecke operators T_p's and U_p's.

If

$$f(z) = \sum_{n=1}^\infty a_n e^{2\pi i n z}$$

is the Fourier expansion at $i\infty$, and $a_1 = 1$, we call it *normalized*. Two eigenforms will be called *equivalent* if they are in the same eigenspace in $S_k(\Gamma_0(N),\epsilon)$ under the action of the Hecke operators.

Theorem 5.2   (Hecke) *The space $S_k(\Gamma_0(N),\epsilon)$ has a basis of normalized eigenfunctions for all T_p's. For each normalized eigenform f,*

$$L(s,f) = \prod_{p|N}\left(1 - \frac{a_p}{p^s}\right)^{-1}\prod_{p\nmid N}\left(1 - \frac{a_p}{p^s} + \frac{\epsilon(p)}{p^{2s+1-k}}\right)^{-1}$$

which converges absolutely for $\operatorname{Re} s > \frac{k+2}{2}$.

Remark.   The product converges for $\operatorname{Re} s > \frac{k+1}{2}$. See [Ogg] for further details.

§6 Oldforms and newforms

Hecke's theorems give no correlation between L-functions having functional equations and those having Euler products. The reason for this difficulty is two-fold. If $d \mid N$, then an element of $S_k(\Gamma_0(d))$ can also be considered as an element of $S_k(\Gamma_0(N))$ and an eigenfunction for all Hecke operators T_p, $p \nmid d$ in $S_k(\Gamma_0(d))$ is also an eigenfunction for all Hecke operators T_p, $p \nmid N$ in $S_k(\Gamma_0(N))$. Also if $f \in S_k(\Gamma_0(d))$, then $f(Nz/d) \in S_k(\Gamma_0(N))$, as a trivial calculation shows. We can combine both of these observations in the general context of $S_k(\Gamma_0(N), \epsilon)$.

Suppose $N' \mid N$ and that ϵ is a Dirichlet character modulo N'. If f is a cusp form on $\Gamma_0(N')$ of type (k, ϵ) and $dN' \mid N$, then $z \mapsto f(dz)$ is a cusp form on $\Gamma_0(N)$ of type (k, ϵ). The forms on $\Gamma_0(N)$ which may be obtained in this way from divisors N' of N, $N' \neq N$, span a subspace $S_k^{old}(\Gamma_0(N), \epsilon)$ called the space of *oldforms*. Its orthogonal complement under the Petersson inner product is denoted $S_k^{new}(\Gamma_0(N), \epsilon)$ and the eigenforms in this space are called *newforms*. We have

$$S_k(\Gamma_0(N), \epsilon) = S_k^{old}(\Gamma_0(N), \epsilon) \bigoplus S_k^{new}(\Gamma_0(N), \epsilon)$$

Theorem 6.1 (Atkin-Lehner) *If f is a newform then its equivalence class is one-dimensional.*

If f is a newform of level N, then $L(s, f)$ extends to an entire function, has an Euler product and satisfies a functional equation. We say that f is of CM type if there is a quadratic field K such that $a(p) = 0$ whenever $p \nmid N$ and p is inert in K. The analytic behaviour of the coefficients of f varies according as f is or is not of CM-type.

§7 The Sato-Tate conjecture

Let

$$f(z) = \sum_{n=1}^{\infty} a(n)e^{2\pi i n z}$$

be a newform of weight k and level N which is also a cusp form. Let us write for each prime $p \nmid N$,

$$a(p) = 2p^{\frac{k-1}{2}} \cos \theta_p.$$

Since we know the Ramanujan conjecture, the θ_p's are real. Inspired by the Sato-Tate conjecture for elliptic curves, Serre [Se] conjectured that if f is not of CM-type, then θ_p's are uniformly distributed with respect to the Sato-Tate measure

$$\frac{2}{\pi} \sin^2 \theta.$$

If we consider the group $SU(2, \mathbb{C})$ and consider its space of conjugacy classes, we can make the following assignment:

$$p \mapsto \text{conjugacy class of } \begin{pmatrix} e^{i\theta_p} & 0 \\ 0 & e^{-i\theta_p} \end{pmatrix}.$$

The Haar measure on the space of conjugacy classes in $SU(2, \mathbb{C})$ is

$$\frac{2}{\pi} \sin^2 \theta.$$

Therefore, following the formalism of the previous chapter, we see that the Sato-Tate conjecture is true if a certain family of L-functions admit an analytic continuation to $\mathrm{Re}(s) \geq 1$ and do not vanish there. More precisely, consider for each $m \geq 1$,

$$L_m(s) = \prod_{p \nmid N} \prod_{j=0}^{m} \left(1 - \frac{e^{i\theta_p(m-2j)}}{p^s} \right)^{-1}.$$

Clearly, $L_m(s)$ converges for $\mathrm{Re}(s) > 1$. It is in fact conjectured that each $L_m(s)$ extends to an entire function. By Theorem III.3.1, we see that the Sato-Tate conjecture is true if and only if $L_m(1 + it) \neq 0$ for every real t and $m \geq 1$. The fact that $L_1(s)$ extends to an entire function follows from Hecke's theory. In 1939, Rankin and Selberg (independently) introduced a powerful method into the theory of numbers and as a consequence established that

$$(s - 1)\zeta(s)L_2(s)$$

extends to an entire function. It was Shimura who using the theory of modular forms of half-integral weight managed to isolate $L_2(s)$ and established an analytic continuation for all $s \in \mathbb{C}$. By the powerful methods of the Langlands program, Shahidi has established the analytic continuation for $L_3(s)$, and $L_4(s)$ to $\mathrm{Re}(s) \geq 1$. (See [Sh].) In some cases, he has obtained better results establishing a meromorphic continuation and defining sets where possible poles may exist. Since we do not need these here, we will not go into these details.

Ogg [Ogg2] has shown that if for each $r \leq 2m$, $L_r(s)$ has an analytic continuation to $\mathrm{Re}(s) > 1/2 - \delta$ for some $\delta > 0$, then $L_m(1 + it) \neq 0$. K. Murty [VKM2] showed that it suffices to have analytic continuation of each $L_m(s)$ up to $\mathrm{Re}(s) \geq 1$.

§8 Oscillations of Fourier coefficients of newforms

Deligne's theorem proving the Ramanujan-Petersson conjecture implies that if f is a newform of weight k, then

$$|a_n| \leq n^{\frac{k-1}{2}} d(n)$$

where $d(n)$ is the divisor function. It is known that the maximum order of $d(n)$ satisfies

$$d(n) = \mathbf{O}(\exp(c \log n / \log \log n))$$

for some constant $c > 0$. Therefore,

$$a_n = \mathbf{O}(n^{\frac{k-1}{2}} \exp\left(\frac{c \log n}{\log \log n}\right)).$$

We would like to know if this is best possible. This question has a long history. Before we discuss the past accomplishments on this question, we recall the Ω notation.

Let g be a positive function and f any function. We say

$$f(x) = \Omega(g(x))$$

if there is some constant $c > 0$ such that $|f(x)| > cg(x)$ for infinitely many $x \to \infty$. We also write

$$f(x) = \Omega_{\pm}(g(x))$$

if there exists a constant $c > 0$ such that

$$f(x) > cg(x)$$

for infinitely many $x \to \infty$ and

$$f(x) < -cg(x)$$

for infinitely many $x \to \infty$.

Hardy proved that

$$a_n = \Omega(n^{\frac{k-1}{2}}).$$

Rankin showed

$$\lim_{n \to \infty} \sup \frac{|a_n|}{n^{(k-1)/2}} = +\infty.$$

Then Joris proved that for $\delta_k = 6/k^2$, we have

$$a_n = \Omega(n^{(k-1)/2} \exp(c(\log n)^{\delta_k - \epsilon})).$$

This was improved by Balasubramanian and R. Murty to $\delta_k = 1/k$. In the special case of the Ramanujan τ-function, they showed

$$\tau(n) = \Omega(n^{11/2} \exp(c(\log n)^{2/3 - \epsilon})).$$

In 1983, R. Murty [RM] proved the following result.

Theorem 8.1   *For any normalized newform f of weight k and level N, there is a constant c > 0 such that*

$$a_n = \Omega_{\pm} \left(n^{\frac{k-1}{2}} \exp\left(\frac{c \log n}{\log \log n} \right) \right).$$

In view of the above discussion, this is best possible apart from the value of the constant $c > 0$.

For any cusp form of weight k, he also established the following result:

Theorem 8.2   *For any cusp form f of weight k and level N, there is a constant c > 0 such that*

$$a_n = \Omega \left(n^{\frac{k-1}{2}} \exp\left(\frac{c \log n}{\log \log n} \right) \right).$$

To prove these, we first need the non-vanishing of $L_3(s)$ and $L_4(s)$ on $\mathrm{Re}(s) = 1$. To do this, we prove a slightly more general theorem:

Theorem 8.3   *If $L_r(s)$ has an analytic continuation up to $\mathrm{Re}(s) \geq 1/2$ for $1 \leq r \leq 2m$, then $L_{2m-1}(1 + it) \neq 0$ for $t \in \mathbb{R}$. If $L_{2r}(s)$ has an analytic continuation up to $\mathrm{Re}(s) = 1$ for $1 \leq r \leq m$, then $L_{2m}(1 + it) \neq 0$.*

Corollary 8.4 $L_3(1 + it) \neq 0$ and $L_4(1 + it) \neq 0$ for all $t \in \mathbb{R}$.

Proof of the Corollary. By the history described at the end of the previous section, we know that $L_1(s)$, $L_2(s)$, $L_3(s)$ and $L_4(s)$ extend to analytic functions for $\mathrm{Re}(s) \geq 1$. The result now follows from the theorem.

Proof of Theorem 8.3. We first show that $L_{2m}(1 + it) \neq 0$. Consider

$$f(s) = L_0 L_2 \cdots L_{2m}.$$

Then, by the trigonometric identities (see exercise)

$$\frac{1}{2} + \cos \theta + \cos 2\theta + \cdots \cos n\theta = \frac{\sin(n + \frac{1}{2})\theta}{2 \sin(\theta/2)}$$

and

$$\cos \theta + \cos 3\theta + \cdots \cos(2n - 1)\theta = \frac{\sin n\theta}{2 \sin \theta}$$

we see that

$$\log L_r(s) = \sum_{n,p} \left(\frac{\sin(r + 1)n\theta_p}{\sin n\theta_p} \right) \frac{1}{np^{ns}}.$$

Therefore, as

$$1 + \frac{\sin 3\theta}{\sin \theta} + \frac{\sin 5\theta}{\sin \theta} + \cdots + \frac{\sin(2n - 1)\theta}{\sin \theta} = \left(\frac{\sin n\theta}{\sin \theta} \right)^2,$$

we find that $\log f(s)$ is a Dirichlet series with non-negative coefficients. Moreover, the Euler product shows that $f(s)$ does not vanish in $\sigma > 1$. An application of Theorem I.1.2 with $e \leq 1$ gives the result. Now consider,

$$g(s) = (L_0 L_1 \cdots L_{2m-1})^2 L_{2m}.$$

An easy computation gives

$$\log g(s) = \sum_{n,p} \left((2m+1) + \sum_{j=0}^{2m-1} 2(j+1) \cos(2m-j)n\theta_p \right) \frac{1}{np^{ns}}.$$

Since

$$2m + 1 + \sum_{j=0}^{\infty} 2(j+1) \cos(2m-j)\theta = \left(\frac{\sin(m+1/2)\theta}{\sin(\theta/2)} \right)^2,$$

we see that $\log g(s)$ is a Dirichlet series with non-negative coefficients. If $L_{2m-1}(1+it) = 0$, with $t \neq 0$, then $g(s)$ has a zero of order ≥ 2 on $\mathrm{Re}(s) = 1$, $s \neq 1$. As $g(s)$ has a pole of order 2 at $s = 1$, we get a contradiction again by Theorem I.1.2. We need to consider $L_{2m-1}(1) = 0$. If this happens, then $g(s)$ is regular. By a well-known theorem of Landau (see exercise 5 in Chapter 1), we find that $\log g(s)$ has a singularity at its abscissa of convergence. As $L_0(s) = \zeta(s)$ has zeros in $\mathrm{Re}(s) \geq 1/2$, $g(s)$ has zeros in this half-plane. Therefore the abscissa of convergence of $\log g(s)$ is $\sigma_0 \geq 1/2$, and as $g(s)$ is analytic in $\mathrm{Re}(s) \geq 1/2$, σ_0 is a zero of $g(s)$. But then $g(\sigma) \geq 1$ for $\sigma > \sigma_0$. We get a contradiction by letting $\sigma \to \sigma_0^+$. This completes the proof of the theorem.

Theorem 8.4 *Suppose $L_r(s)$ has an analytic continuation up to $\mathrm{Re}(s) \geq 1/2$ for all $r \leq 2m+2$. Then*

(i) *for $r \leq m+1$,*

$$\sum_{p \leq x} (2 \cos \theta_p)^{2r} = \frac{1}{r+1} \binom{2r}{r} (1 + o(1)) \frac{x}{\log x}$$

as $x \to \infty$, and

(ii) *for $r \leq m$,*

$$\sum_{p \leq x} (2 \cos \theta_p)^{2r+1} = o(x/\log x)$$

as $x \to \infty$.

Proof. By Theorem 8.3, we know that $L_r(s)$ does not vanish on the line $\sigma = 1$. Observe further that

$$\frac{\sin(r+1)\theta_p}{\sin \theta_p} = \sum_{j=0}^{r} e^{i(r-2j)\theta}.$$

Therefore, by the Tauberian theorem, we deduce for $1 \le r \le 2m+2$,

$$\sum_{p \le x} \frac{\sin(r+1)\theta_p}{\sin \theta_p} = \mathbf{o}(x/\log x)$$

as $x \to \infty$. Writing $U_n(\cos \theta) = \frac{\sin(n+1)\theta}{\sin \theta}$ and $T_n(\cos \theta) = \cos n\theta$ for each $n \ge 1$, we find for $2 \le r \le 2m+2$,

$$\sum_{p \le x} T_r(\cos \theta_p) = \mathbf{o}(x/\log x),$$

because of the identity

$$2T_n(x) = U_n(x) - U_{n-2}(x).$$

Note that $T_n(x)$ and $U_n(x)$ are the familiar Chebycheff polynomials. Also,

$$\sum_{p \le x} T_2(\cos \theta_p) = (-\frac{1}{2} + \mathbf{o}(1))\frac{x}{\log x},$$

and

$$\sum_{p \le x} T_1(\cos \theta_p) = \mathbf{o}(x/\log x),$$

as $L_1(s)$ is regular and non-vanishing for $\mathrm{Re}(s) \ge 1$.

Now define $T_0(x) = 1/2$. Then the inverse relation for the Chebycheff polynomials gives

$$(2\cos \theta)^r = 2 \sum_{k=0}^{r'} \binom{r}{k} T_{r-2k}(\cos \theta)$$

where $r' = [r/2]$ (see the exercises). Therefore,

$$\sum_{p \le x} (2\cos \theta_p)^r = 2 \sum_{k=0}^{r'} \sum_{p \le x} T_{r-2k}(\cos \theta_p).$$

By the above results, the inner sum is $\mathbf{o}(x/\log x)$ unless $r - 2k = 2$ or 0. Hence, (ii) is deduced. In the case of (i), we find that

$$\sum_{p \le x} (2\cos \theta_p)^{2r} = \left(-\binom{2r}{r-1} + \binom{2r}{r}\right)(1 + \mathbf{o}(1))\frac{x}{\log x}$$

as $x \to \infty$. The term in the brackets is easily seen to be the coefficient stated in (i).

To prove the omega theorem, we will need a few preliminary combinatorial identities. We collect them below and leave them as exercises.

Lemma 8.5

(i)
$$\sum_{j=0}^{r}(-1)^j \binom{r}{j}\binom{2j}{j}\frac{2^{-2j}}{j+1} = 2^{-2r-1}\binom{2r+2}{r+1},$$

(ii)
$$\sum_{j=0}^{r}(-1)^j \binom{r}{j}\binom{2j+2}{j+1}\binom{2j+2}{j+1}\frac{2^{-2j}}{j+2} = \frac{2^{-2r}}{r+2}\binom{2r+2}{r+1}.$$

Proof. Exercise.

Theorem 8.6 *Suppose that $L_r(s)$ has an analytic continuation up to $\mathrm{Re}(s) \geq 1/2$ for all $r \leq 2m+2$. Then, each of the statements holds for a set of primes of positive density:*

(i) *for any $\delta > 0$,*
$$-\delta < 2\cos\theta_p < \frac{2}{\delta(m+2)},$$

(ii) *for any $\epsilon > 0$,*
$$|2\cos\theta_p| > \sqrt{\frac{4m+2}{m+2}} - \epsilon,$$

(iii) *for any $\epsilon > 0$, $2\cos\theta_p > \beta_m - \epsilon$, where $\beta_m = \left\{\frac{1}{4(m+2)}\binom{2m+2}{m+1}\right\}^{\frac{1}{2m+1}}$.*
There is a corresponding result for negative values of a_p.

Proof. For (i), consider the polynomial

$$P_m(x) = (x^2 - 4)^m(x-\alpha)(x-\beta),$$

where α, β will be suitably chosen later. By Theorem 8.4 and Lemma 8.5, we deduce

$$\frac{\log x}{x}\sum_{p\leq x} P_m(2\cos\theta_p) \sim (-1)^m\binom{2m+2}{m+1}\left(\frac{\alpha\beta}{2} + \frac{1}{m+2}\right).$$

Examining the graph of $P_m(x)$ and choosing α, β so that

$$\alpha\beta > -\frac{2}{m+2}, \quad \text{if } m \text{ is even and} \quad \alpha\beta < -\frac{2}{m+2},$$

if m is odd, we set $\alpha = -\delta$ to get the desired result. To prove (ii), consider

$$Q_m(x) = x^{2m}(x^2 - \gamma)$$

where γ shall soon be chosen. By Theorem 8.4,

$$\frac{\log x}{x}\sum_{p\leq x} Q_m(2\cos\theta_p) \sim \frac{1}{m+2}\binom{2m+2}{m+1} - \gamma\frac{1}{m+1}\binom{2m}{m}.$$

Examining the graph of $Q_m(x)$, we note that if

$$\gamma = \frac{m+1}{m+2}\binom{2m+2}{m+1}\binom{2m}{m}^{-1} = \frac{4m+2}{m+2},$$

we obtain (ii). To prove (iii), we begin by noting

$$\sum_{p \le x} |2\cos\theta_p|^{2m+1} \ge \frac{1}{2}\sum_{p \le x}(2\cos\theta_p)^{2m+2}.$$

By Theorem 8.4, we find

$$4\sum_{\substack{p \le x \\ a(p) > 0}}(2\cos\theta_p)^{2m+1} \gtrsim \frac{1}{m+2}\binom{2m+2}{m+1}\frac{x}{\log x}$$

as $x \to \infty$. Thus, for a positive proportion of the primes

$$2\cos\theta_p > \left\{\frac{1}{4(m+2)}\binom{2m+2}{m+1}\right\}^{\frac{1}{2m+1}} - \epsilon.$$

This completes the proof of the theorem.

The proofs of Theorem 8.1 and 8.2 are now immediate. We relegate these to the exercises below. One should consult the appendix due to Serre in [Sh] for certain improvements of these results.

§9 Rankin's theorem

Rankin [R] proved the following theorem:

Theorem 9.1 *Let f be a normalized newform which is a cusp form of weight k with respect to $\Gamma_0(N)$. Let a_n denote the n-th Fourier coefficient and write*

$$a(n) = a_n / n^{(k-1)/2}.$$

Given $\beta \ge 0$, let

$$F(\beta) = \frac{2^{\beta-1}}{5}(2^\beta + 3^{2-\beta}) - 1.$$

Then

$$\sum_{n \le x} |a(n)|^{2\beta} \ll x(\log x)^{F(\beta)}.$$

In particular, if $k = 2$, then

$$\sum_{n \le x} |a(n)| \ll x(\log x)^{-1/18}.$$

The proof of Rankin's theorem can be found in [R] or [Sh, p. 174–175]. It relies in a crucial way on the analytic continuation of $L_4(s)$ to $\text{Re}(s) \ge 1$.

Exercises

1. Show that
$$[SL_2(\mathbb{Z}) : \Gamma(N)] = N^3 \prod_{\substack{p|N \\ p \text{ prime}}} \left(1 - \frac{1}{p^2}\right)$$

2. Show that:
   (i) $\Gamma_0(N)$ is not a normal subgroup of $SL_2(\mathbb{Z})$ if $N > 1$.
   (ii) $\Gamma_0(N)$ has index $N \prod_{p|N} \left(1 + \frac{1}{p}\right)$ in $SL_2(\mathbb{Z})$.
   (iii) $\Gamma_1(N)$ is not normal in $SL_2(\mathbb{Z})$ but is normal in $\Gamma_0(N)$. Compute its index.
   (iv) $\gamma \in SL_2(\mathbb{Z})$ has finite order if and only if γ is elliptic.

3. Show that dim $S_2(2) = 0$.

4. Assuming the Sato-Tate conjecture, determine the largest positive constant c such that the estimate (in the notation of Section 9),
$$\sum_{n \leq x} |a(n)| \ll x(\log x)^{-c}$$
is valid for all sufficiently large x.

5. Show that if $f \in S_k(\Gamma_0(N), \epsilon)$, then $f|W_N \in S_k(\Gamma_0(N), \bar{\epsilon})$. Derive the functional equation for $L(s, f)$.

6. Using the fact that $2\cos\theta_p > \sqrt{2} - \epsilon$ for a positive proportion of primes deduce that
$$a_n = \Omega(n^{(k-1)/2} \exp(c \log n / \log \log n)),$$
for some positive constant c. Deduce a corresponding result for negative oscillation.

7. Given any cusp form f of level 1, show that there are numbers m_i which are not all zero so that
$$\sum_{i=1}^{r} m_i T_i(f)$$
is an eigenform. Here, T_i denotes the i-th Hecke operator and r is the dimension of the space of cusp forms of weight k. Now deduce Theorem 8.2 for any cusp form of level 1.

References

[CO] H. Cohen and J. Oesterlé, Dimensions des espaces de formes modulaires, Lecture Notes in Mathematics, 627(1977), 69–78.

[D] W. Duke, The dimension of the space of cusp forms of weight one, Int. Math. Res. Not., 1995, pp. 99–109.

[F] G. Frey, Construction and arithmetical application of modular forms of low weight, in: Elliptic curves and related objects, pp. 1–21, eds. H. Kisilevsky and R. Murty, CRM Proceedings and Lecture Notes, vol. 4, Amer. Math. Soc., Providence, 1994.

[Kn] A. Knapp, Elliptic Curves, Princeton University Press, 1992.

[La] S. Lang, Introduction to modular forms, Grundl. Math. Wiss. **222**, Springer-Verlag, Berlin, New York 1976.

[Ogg] A. Ogg, Survey of modular functions of one variable, in: 'Modular functions of one variable I', (Ed. W. Kuijk), Lecture Notes in Math., **320** (1972) pp. 1–36, Springer-Verlag.

[Ogg2] A. Ogg, A remark on the Sato-Tate conjecture, Inventiones Math., **9** (1970) p. 198–200.

[RM] M. Ram Murty, Oscillations of Fourier coefficients of modular forms, Math. Annalen, **262** (1983) p. 431–446.

[R] R. Rankin, Sums of powers of cusp form coefficients II, Math. Ann., **272** (1985) p. 593–600.

[VKM] V. Kumar Murty, Introduction to abelian varieties, CRM Monograph Series, Volume **3**, 1993, American Math. Society, Providence, USA. Chapters 9–11.

[VKM2] V. Kumar Murty, On the Sato-Tate conjecture, in: Number Theory related to Fermat's Last Theorem, (ed. N. Koblitz), Birkhäuser-Verlag, Boston, 1982, pp. 195–205.

[Sh] F. Shahidi, Symmetric power L-functions for $GL(2)$, in: Elliptic curves and related objects, (ed. H. Kisilevsky and M. Ram Murty), CRM Proceedings and Lecture Notes, Vol. 4 (1994) pp. 159–182.

[Shi] G. Shimura, Introduction to the arithmetic theory of automorphic functions, Publ. Math. Soc. Japan, vol. 11, Iwanami Shoten, 1977.

Chapter 5
Dirichlet L-Functions

§1 Introduction

Let χ denote a Dirichlet character and $L(s, \chi)$ the associated Dirichlet L-function. Let us begin by considering how one would approach the problem of showing that $L(\frac{1}{2}, \chi) \neq 0$. In the following, we assume that χ is defined modulo a prime q. We first study the average

$$\sum_{\chi \,(\text{mod } q)} L(\tfrac{1}{2}, \chi).$$

By the approximate functional equation, one can show that

$$L(\tfrac{1}{2}, \chi) \sim \sum_{n < \sqrt{q}} \frac{\chi(n)}{\sqrt{n}}.$$

Hence,

$$\sum_{\chi \,(\text{mod } q)} L(\tfrac{1}{2}, \chi) \sim \sum_{\chi} \sum_{n < \sqrt{q}} \frac{\chi(n)}{\sqrt{n}} \sim \phi(q).$$

Similarly,

$$\sum_{\chi \,(\text{mod } q)} |L(\tfrac{1}{2}, \chi)|^2 \sim \sum_{\chi} \sum_{n_1, n_2 < q} \frac{\chi(n_1)\bar{\chi}(n_2)}{\sqrt{n_1 n_2}}$$

$$\sim \phi(q) \sum_{n < q} \frac{1}{n}$$

$$\sim \phi(q) \log q.$$

Therefore,

$$\phi(q)^2 \ll \left(\sum L(\tfrac{1}{2}, \chi)\right)^2$$

$$\ll \left(\sum |L(\tfrac{1}{2}, \chi)|^2\right) \left(\sum_{L(\frac{1}{2}, \chi) \neq 0} 1\right)$$

$$\ll \phi(q)(\log q) \left(\sum_{L(\frac{1}{2}, \chi) \neq 0} 1\right)$$

and so

$$\#\{\chi \bmod q : L(\tfrac{1}{2}, \chi) \neq 0\} \gg \frac{\phi(q)}{\log q}.$$

This kind of argument can in fact be made precise. The reader is referred to the paper [B] of Balasubramanian where only a slightly weaker result is proved.

We may try to consider higher moments as well. In general, one might expect that for some $c > 0$ depending on k,

$$\sum |L(\tfrac{1}{2}, \chi)|^{2k} \sim c\phi(q) \sum_{r < q^k} \frac{d_k(r)^2}{r}$$

where $d_k(r)$ denotes the number of decompositions of r as a product of k positive integers. Now, $d_k(p) = k$ so the Dirichlet series

$$\sum \frac{d_k(r)^2}{r^s}$$

has Euler product

$$\prod_p \left(1 + \frac{k^2}{p^s} + \cdots\right)$$

which near $s = 1$ behaves like

$$\zeta(s)^{k^2} \sim (s - 1)^{-k^2}.$$

Thus for some $c' > 0$, depending on k,

$$\sum_{r < X} d_k(r)^2 \sim c' X (\log X)^{k^2 - 1}$$

and so we might expect

$$\sum |L(\tfrac{1}{2}, \chi)|^{2k} \sim c_k \phi(q)(\log q)^{k^2}. \qquad (*)$$

If we assume this and try to apply the above argument, we do not seem to get anything better. Of course, (∗) is still very highly conjectural. The largest value for which an asymptotic formula is known is $k = 2$, and prime q. This is a result of Heath-Brown [HB].

In spite of our difficulty in showing that $L(\frac{1}{2}, \chi) \neq 0$ holds often, no example is known for which $L(\frac{1}{2}, \chi) = 0$. Siegel [Sie] has shown that any point on the line $\sigma = \frac{1}{2}$ is a limit point of zeroes of the $L(s, \chi)$ as χ ranges over *all* Dirichlet characters. More precisely, he shows that for any $[T_1, T_2]$, with $T_2 - T_1 \gg (\log q)^{-\frac{1}{2}}$, there exists a $\chi(\bmod q)$ with $L(s, \chi) = 0$ at some point $s = \sigma + it$ in the rectangle

$$\sigma \in [\frac{1}{2}, \frac{1}{2} + \frac{1}{\log\log q}], \quad t \in [T_1, T_2].$$

The dimensions of this rectangle can be somewhat reduced now.

A variant of these problems is to restrict attention to real characters χ. In this case, one can average over such characters only and show that many of them do not vanish.

In this chapter, we shall look at both of these problems. After establishing the fundamental estimate of Polya and Vinogradov in Section 2, we discuss in §§3–4 the work of Jutila which establishes an asymptotic formula for the sum

$$L(\frac{1}{2}, \chi_D)$$

where χ_D is the real character given by $(\frac{D}{\cdot})$. From this, one can deduce the non-vanishing of infinitely many such L-functions. In §5, we describe the work of R. Balasubramanian and V. K. Murty [BM] which considers Dirichlet L-functions to a prime modulus q. Here, it is proved that for a positive proportion of the $\chi(\bmod q)$, we have $L(\frac{1}{2}, \chi) \neq 0$. Finally in §6, we briefly describe the work of R. Murty [RM] which establishes a strengthening of §5 assuming the Riemann Hypothesis .

§2 Polya-Vinogradov estimate

In this section, we shall describe an estimate for character sums involving a Dirichlet character. Let $q \geq 1$ be an integer and let χ denote a character $\bmod q$. We shall consider a sum of the form

$$\sum_{n \leq x} \chi(n)$$

or more generally,

$$\sum_{y \leq n \leq x} \chi(n).$$

Since χ is periodic mod q, we certainly have the trivial estimate

$$\sum_{y \leq n \leq x} \chi(n) \leq \min(q, x - y + 1).$$

If χ is the trivial character, then this is best possible. However, for nontrivial χ it is possible to do substantially better.

Theorem 2.1 *We have the estimate*

$$\sum_{y < n \leq x} \chi(n) \ll q^{\frac{1}{2}} \log q.$$

Remark. If χ is primitive, then the implied constant may be taken to be one. In general, sharp estimates for the constant, as well as connections of this problem with diophantine approximation have been given in work of Hildebrand [Hil].

Proof. First, suppose that χ is primitive. We have the identity in terms of Gauss sums:

$$\chi(n) = \frac{1}{g(\bar{\chi})} \sum_{a=1}^{q} \bar{\chi}(a) e\left(\frac{an}{q}\right)$$

where $e(\alpha) = \exp(2\pi i \alpha)$. Summing both sides over n, we have to estimate

$$\frac{1}{g(\bar{\chi})} \sum_{a=1}^{q} \bar{\chi}(a) \sum_{y < n \leq x} e\left(\frac{an}{q}\right).$$

Since the inner sum is a geometric series, it can be evaluated exactly and we find it to be

$$e\left(\frac{(M + \frac{1}{2}N + \frac{1}{2})a}{q}\right) \frac{\sin \pi N a/q}{\sin \pi a/q}$$

where $M = [y] + 1$ and $N = [x] - [y] - 1$. Hence, using the fact that the Gauss sum $g(\bar{\chi})$ has absolute value $q^{\frac{1}{2}}$ we deduce that

$$\left| \sum_{y \leq n \leq x} \chi(n) \right| \leq q^{-\frac{1}{2}} \sum_{a=1}^{q-1} \frac{1}{\sin \pi a/q}.$$

Using the estimate

$$\sin \pi \beta > 2\beta \quad \text{if } 0 < \beta < \frac{1}{2}$$

we deduce that

$$\sum_{a=1}^{q-1} \frac{1}{\sin \pi a/q} < 1 + q \sum_{a=1}^{[q/2]} \frac{1}{a} < 1 + q \log q/2.$$

Thus

$$\left| \sum_{y \leq n \leq x} \chi(n) \right| < q^{\frac{1}{2}} \log q.$$

The proof of the estimate in the case that χ is not primitive is left as an exercise.

One can ask whether the estimate provided by the Polya-Vinogradov inequality is best possible. Montgomery and Vaughan [MV] have shown that assuming the Riemann Hypothesis for Dirichlet L-functions, we have the sharper estimate

$$\sum_{n \leq x} \chi(n) \ll q^{\frac{1}{2}} \log \log q.$$

It is possible to remove the factor $\log q$ entirely if we work with even characters and integrate over the summation parameter. For a character of any parity, it is possible to eliminate the $\log q$ factor if we integrate *and* take the mean square. These observations play a crucial role in Chapter 6.

§3 Jutila's character sum estimate

The character sum referred to is the following:

$$S(X, Y) = \sideset{}{'}\sum_{|D| \leq X} \left| \sum_{n \leq Y} \left(\frac{D}{n} \right) \right|^2$$

where the outer sum is over D which belong to the set $\mathcal{D}$ of integers satisfying: (i) D is not a square and (ii) $D \equiv 1 \pmod 4$ or $D = 4N$ with $N \not\equiv 0 \pmod 4$. Note that if $(D/\cdot)$ is a non-principal character, then $D \in \mathcal{D}$.

If we apply the Polya-Vinogradov estimate to the inner sum, we find that it is

$$\ll \sqrt{|D|} (\log |D|).$$

From this, we deduce that

$$S(X, Y) \ll X^2 (\log X)^2.$$

In [Jut1], Jutila proves the following.

Theorem 3.1 *For $X \geq 3$ and $Y \geq 1$, we have*

$$S(X, Y) \ll XY (\log X)^2.$$

Remark. In [Jut1], one actually finds the weaker estimate $XY (\log X)^8$. Several authors, including Jutila (see [Jut2, p. 155]), observed that the methods of [Jut1] in fact yield Theorem 3.1.

The main tool in the proof of Theorem 3.1 is a Lemma of Vinogradov [Vino] which gives an explicit and effective smooth approximation to the characteristic function of an interval. We state the lemma as follows.

Lemma 3.2 *Let α, β and δ be real numbers satisfying*

$$0 < \delta < \frac{1}{2}, \ \delta \le \beta - \alpha \le 1 - \delta.$$

Then there exists a periodic function $\psi(x)$ with period 1, and satisfying
(i)   *$\psi(x) = 1$ in the interval $[\alpha + \frac{1}{2}\delta, \beta - \frac{1}{2}\delta]$*
(ii) *$\psi(x) = 0$ in the interval $[\beta + \frac{1}{2}\delta, 1 + \alpha - \frac{1}{2}\delta]$*
(iii) *$0 \le \psi(x) \le 1$ in the intervals $[\alpha - \frac{1}{2}\delta, \alpha + \frac{1}{2}\delta]$ and $[\beta - \frac{1}{2}\delta, \beta + \frac{1}{2}\delta]$*
(iv) *$\psi(x)$ has the Fourier expansion*

$$\psi(x) = \beta - \alpha + \sum_{m=1}^{\infty} (a_m \mathbf{e}(mx) + \bar{a}_m \mathbf{e}(-mx))$$

where

$$a_m = (2\pi i m)^{-1} (\mathbf{e}(-m\alpha) - \mathbf{e}(-m\beta)) \{ \frac{\sin(\frac{1}{2}\pi m \delta)}{\frac{1}{2}\pi m \delta} \}^2$$

and

$$|a_m| \ll \min(\beta - \alpha, m^{-1}, \delta^{-2} m^{-3}).$$

In the proof of Jutila's estimate, one uses not a single function ψ but a family of such functions. Denote by $\psi(x, u)$ the function $\psi(x)$ of Lemma 3.2 with the parameters

$$\delta = X^{-\frac{1}{2}}, \ \alpha = \frac{1}{2}\delta, \ \beta = Yu^{-1} + \frac{1}{2}\delta.$$

Then one has the estimates

$$\frac{\partial \psi(x, u)}{\partial x} \ll \delta^{-1}$$

and

$$\frac{d\psi(n/u, u)}{du} \ll \frac{X}{u^2}.$$

Moreover

$$\frac{da_{m,u}}{du} \ll \frac{Y}{u^2} \min(1, m^{-2}\delta^{-2}).$$

Proof of Theorem 3.1. First, we observe that the estimate is easy in certain ranges of the parameters. For example, the case $Y \geq X/4$ follows from the Polya-Vinogradov estimate and the case $Y \leq X^{\frac{1}{2}}$ also follows easily and is left as an exercise. We are now left with the essential range $X^{\frac{1}{2}} \leq Y \leq X/4$. Moreover, the Polya-Vinogradov estimate also implies that

$$S(\sqrt{XY}, Y) \ll XY (\log XY)^2.$$

Hence, we may assume in the summation over $|D|$ that

$$Y \leq \sqrt{XY} \leq |D| \leq X.$$

Using the fact that

$$\psi(\frac{n}{|D|}, |D|) = \begin{cases} 1 & \text{if } \frac{|D|}{\sqrt{X}} \leq n \leq Y \\ 0 & \text{if } Y + \frac{|D|}{\sqrt{X}} \leq n \leq |D| \end{cases}$$

and the above inequality on $|D|$, we see that

$$\sum_{n \leq |D|} \psi(\frac{n}{|D|}, |D|) \left(\frac{D}{n}\right) = \sum_{n \leq X^{\frac{1}{2}}} \psi(\frac{n}{|D|}, |D|) \left(\frac{D}{n}\right) + \sum_{\sqrt{X} < n \leq Y} \left(\frac{D}{n}\right)$$

$$- \sum_{Y < n < Y + |D|/\sqrt{X}} \psi(\frac{n}{|D|}, |D|) \left(\frac{D}{n}\right).$$

Using the fact that

$$Y + \frac{|D|}{\sqrt{X}} \leq Y + \sqrt{X} < |D|$$

the last sum can be extended to the range $Y < n < Y + \sqrt{X}$. Let us write

$$S(D) = \sum_{n \leq Y} \left(\frac{D}{n}\right).$$

Then, we can decompose the inner sum as follows:

$$S(D) = S_1(D) + S_2(D) + S_3(D)$$

where

$$S_1(D) = \sum_{n \leq |D|} \psi(\frac{n}{|D|}, |D|) \left(\frac{D}{n}\right)$$

$$S_2(D) = \sum_{n \leq X^{\frac{1}{2}}} (1 - \psi(\frac{n}{|D|}, |D|)) \left(\frac{D}{n}\right)$$

and

$$S_3(D) = \sum_{Y \leq n \leq Y + X^{\frac{1}{2}}} \psi(\frac{n}{|D|}, |D|) \left(\frac{D}{n}\right).$$

We consider the sum

$$S = \sum_{\sqrt{XY}<|D|\leq X}^{*} |S(D)|^2 \ll \sum_{i=1}^{3} \sum_{\sqrt{XY}<|D|\leq X}^{*} |S_i(D)|^2$$

where $*$ denotes a sum over fundamental discriminants.

Now, we can use Vinogradov's Lemma to estimate these sums. Suppose now that D is a fundamental discriminant. First, we have

$$S_1(D) = g_D \sum_{m=1}^{\infty} \left(a_{m,|D|}\left(\frac{D}{m}\right) + \bar{a}_{m,|D|}\left(\frac{D}{-m}\right)\right)$$

where $g_D = g\left(\left(\frac{D}{\cdot}\right)\right)$ is the Gauss sum. Let us split the sum over m into two segments $S_{11}(D)$ and $S_{12}(D)$ depending on $m \leq X^{\frac{1}{2}}$ and $m > X^{\frac{1}{2}}$. We have

$$\sum_{\sqrt{XY}<|D|\leq X}^{*} |S_{11}(D)|^2 \ll \sum_{\sqrt{XY}<|D|\leq X}^{'} |D| \left| \sum_{m\leq X^{\frac{1}{2}}} a_{m,|D|}\left(\frac{D}{m}\right)\right|^2$$

which on expanding is equal to

$$\sum_{r,s\leq X^{\frac{1}{2}}} \sum_{\sqrt{XY}<|D|\leq X}^{'} |D| a_{r,|D|}\bar{a}_{s,|D|}\left(\frac{D}{rs}\right). \tag{3.1}$$

To estimate this, first consider the case when rs is a square. Using the estimates provided by Vinogradov's lemma, we have to consider

$$2 \sum_{\substack{r<s\leq X^{\frac{1}{2}} \\ rs=t^2}} \sum_{|D|\leq X} |D| \min\frac{Y}{|D|}\frac{1}{r} \min\frac{Y}{|D|}\frac{1}{s}.$$

The inner sum over D is split into three subsums depending on the value of the minima, namely $sY < |D| \leq X$, $rY < |D| \leq sY$ and $|D| \leq rY$. Consider the first range. The sum is

$$\ll \sum_{\substack{r<s\leq X^{\frac{1}{2}} \\ rs=t^2}} \sum_{sY<|D|\leq X} |D|\frac{Y^2}{|D|^2}$$

$$\ll Y^2 \sum_{\substack{r<s<X/Y \\ rs=t^2}} \left(\log\left(\frac{X}{Ys}\right) + \mathbf{O}(\frac{1}{Ys})\right) \tag{3.2}$$

$$\ll Y^2 \sum_{s\leq X/Y} \log\left(\frac{X}{Ys}\right) \sum_{\substack{t<s \\ s_0|t}} 1 + \mathbf{O}(Y^2 \sum_{s\leq X/Y} \frac{1}{Ys}).$$

Here s_0 is defined as follows: if $p^a||s$ then $p^b||s_0$ where b is the least integer $\geq a/2$. It is easy to show that

$$\sum_{n=1}^{\infty} \frac{n/n_0}{n^s} = \zeta(2s-1)\zeta(s)/\zeta(2s).$$

Hence,

$$\sum_{n\leq x} n/n_0 \ll x(\log x).$$

Inserting this into (3.2), using partial summation, and observing that the **O** term is $\mathbf{O}(Y\log X)$, we deduce that (3.2) is

$$\ll Y^2 \sum_{s\leq X/Y} \frac{s}{s_0}\log\frac{X}{Ys} \ll XY(\log X/Y).$$

For the sum in the range $rY < |D| < sY$ we have an estimate

$$\ll \sum_{\substack{r<s\leq X^{\frac{1}{2}} \\ rs=t^2}} \sum_{|D|\leq X} |D|\frac{Y}{|D|}\frac{1}{s} \ll XY \sum_{\substack{r<s<X^{\frac{1}{2}} \\ rs=t^2}} \frac{1}{s} \ll XY \sum_{s<X^{\frac{1}{2}}} \frac{1}{s}\frac{s}{s_0}$$

$$\ll XY\log X$$

by estimates similar to the ones described above. Finally, in the remaining range $|D| \leq rY$, we have that the sum is

$$\ll \sum_{|D|\leq X} |D| \sum_{\frac{|D|}{Y}<t<X^{\frac{1}{2}}} \frac{1}{t^2} \sum_{rs=t^2} 1 \ll \sum_{|D|\leq X} |D|\frac{Y(\log X)^2}{|D|}$$

$$\ll XY(\log X)^2.$$

Now we estimate the contribution of terms with rs not a perfect square. We introduce the functions

$$f_{r,s}(u) = ua_{r,u}\bar{a}_{s,u} \quad \text{and} \quad g_{r,s}(u) = \sum_{|D|\leq u}{}' \left(\frac{D}{rs}\right).$$

Let $p(n) = 0$ if n is a square and $p(n) = 1$ otherwise. Then the part in consideration is by partial summation

$$\sum_{r,s\leq X^{\frac{1}{2}}} p(rs)\left\{g_{r,s}(X)f_{r,s}(X) - \int_{\sqrt{XY}}^{X} g_{r,s}(u)f'_{r,s}(u)du\right\}.$$

Using the estimates

$$f_{r,s}(X) \ll Xr^{-1}s^{-1}, \quad g_{r,s}(u) \ll (rs)^{\frac{1}{2}}\log X \quad \text{and} \quad f'_{r,s}(u) \ll \frac{Y}{u}(r^{-1}+s^{-1})$$

we find that the sum in question is

$$\ll \sum_{r,s<X^{\frac{1}{2}}} \left(X(\log X)(rs)^{-\frac{1}{2}} + (rs)^{\frac{1}{2}}(\log X)Y(\log X/Y)(r^{-1}+s^{-1}) \right)$$

$$\ll X^{3/2}\log X + XY(\log X)(\log X/Y)$$

$$\ll XY(\log X)^2.$$

Next to estimate S_{12} we proceed in an analogous fashion. Now r,s run over the range $r,s > X^{\frac{1}{2}}$. Explicitly, we are trying to estimate

$$\sum_{r,s>\sqrt{X}} \sideset{}{'}\sum_{\sqrt{XY}\le|D|\le X} |D|a_{r,|D|}\bar{a}_{s,|D|}\left(\frac{D}{rs}\right). \tag{3.3}$$

The pairs for which rs is a square give a contribution

$$\ll X^2 \sum_{t^2>X} \mathbf{d}(t^2)X^2n^{-6} \ll X^{3/2}(\log X)^2.$$

When rs is not a square, the estimates above for $f_{r,s}$ and $f'_{r,s}$ are replaced by

$$f_{r,s}(X) \ll X^3(rs)^{-3}$$

and

$$f'_{r,s}(u) \ll \frac{X^2Y}{ur^2s^2}(r^{-1}+s^{-1}) + X^2(rs)^{-3}.$$

Inserting these, one finds that the contribution of S_{12} is

$$\sum_{r,s>\sqrt{X}} (rs)^{1/2}(\log X)(X^3(rs)^{-3} + X^2Y(\log X/Y)(rs)^{-2}(r^{-1}+s^{-1}))$$

and this is

$$\ll X^{3/2}\log X + XY(\log X)(\log X/Y) \ll XY(\log X)^2.$$

Next we consider the sums S_2 and S_3. The sum S_3 is

$$\sum_{Y<r,s<Y+X^{\frac{1}{2}}} \sideset{}{'}\sum_{|D|\le X} \psi(\frac{r}{|D|},|D|)\psi(\frac{s}{|D|},|D|)\left(\frac{D}{rs}\right).$$

The contribution of pairs r, s with rs a square is estimated by bounding the ψ values by 1 giving a total estimate of

$$\ll X \sum_{\substack{Y<r,s<Y+X^{\frac{1}{2}} \\ rs=t^2}} 1 \ll X \sum_{t\leq Y+X^{\frac{1}{2}}} d(t^2)$$

$$\ll XY (\log X)^2.$$

For the nonsquare terms, we have by partial summation that the sum is

$$\sum_{Y<r,s<Y+X^{\frac{1}{2}}} p(rs)\left\{\psi(\frac{r}{X},X)\psi(\frac{s}{X},X)g_{r,s}(X) - \int_{\sqrt{XY}}^{X} g_{r,s}(u)d\left(\psi(\frac{r}{u},u)\psi(\frac{s}{u},u)\right)\right\}.$$

Now

$$\psi(\frac{s}{u},u) = 0 \quad \text{if } u < (s-Y)\sqrt{X}.$$

By the estimate

$$\frac{d\psi(\frac{n}{u},u)}{du} \ll \frac{X}{u^2}$$

and the Polya-Vinogradov estimate, we see that

$$\int_{\sqrt{XY}}^{X} g_{r,s}(u)d\left(\psi(\frac{r}{u},u)\psi(\frac{s}{u},u)\right) \ll \sqrt{rs}\log X \int_{\max((s-Y)\sqrt{X},\sqrt{XY})}^{X} \frac{X}{u^2}du.$$

Inserting this into the sum over r, s we find that it is

$$\sum_{Y<r<Y+\sqrt{X}} \sqrt{r}\sqrt{X}\log X \sum_{Y<s<Y+\sqrt{X}} \frac{\sqrt{s}}{\max(s-Y,\sqrt{Y})}$$

and this is

$$\ll \sum_{Y<r<Y+\sqrt{X}} \sqrt{r}\sqrt{X}\log X \left\{\sum_{Y<s<Y+\sqrt{Y}} \frac{\sqrt{s}}{\sqrt{Y}} + \sum_{Y+\sqrt{Y}<s<Y+\sqrt{X}} \frac{\sqrt{s}}{s-Y}\right\}$$

$$\ll XY(\log X)^2.$$

A similar estimate holds for S_2.

To complete the proof of the Theorem, we have to estimate the contribution of imprimitive characters (that is, the contribution from those D which are not fundamental discriminants). If $\left(\frac{d}{\cdot}\right)$ is the primitive character inducing $\left(\frac{D}{\cdot}\right)$ and $D = de^2$ we have

$$| \sum_{n \leq Y} \left(\frac{D}{n}\right)|^2 = | \sum_{\substack{n \leq Y \\ (n,e)=1}} \left(\frac{d}{n}\right)|^2$$

$$= | \sum_{\delta | e} \mu(\delta) \left(\frac{d}{\delta}\right) \sum_{m \leq Y\delta^{-1}} \left(\frac{d}{m}\right)|^2$$

$$\leq \mathbf{d}(e) \sum_{\delta | e} | \sum_{n \leq Y\delta^{-1}} \left(\frac{d}{n}\right)|^2.$$

Hence, for fixed values of d and δ with $|d|\delta \leq X$ we have a contribution to $S(X,Y)$ an amount which is

$$\leq | \sum_{n \leq Y\delta^{-1}} \left(\frac{d}{n}\right)|^2 \sum_{\substack{e^2 \leq X|d|^{-1} \\ \delta | e}} \mathbf{d}(e)| \sum_{n \leq Y\delta^{-1}} \left(\frac{d}{n}\right)|^2 \mathbf{d}(\delta) \frac{\sqrt{X}}{\sqrt{|d|\delta}} \log \frac{\sqrt{X}}{\sqrt{|d|\delta}}.$$

Now the summation over $|d| \leq X\delta^{-1}$ for a fixed δ gives a contribution

$$\mathbf{d}(\delta) \frac{\sqrt{X}}{\delta} \sum_{|d| \leq X/\delta} \frac{1}{\sqrt{|d|}} \left| \sum_{n \leq Y/\delta} \left(\frac{d}{n}\right) \right|^2 \log \frac{\sqrt{X}}{\sqrt{|d|\delta}}.$$

Applying the result for fundamental discriminants and using partial summation, we deduce that this is

$$\ll \mathbf{d}(\delta) \frac{\sqrt{X}}{\delta} \sqrt{\frac{X}{\delta} \frac{Y}{\delta}} (\log \frac{X}{\delta})^2.$$

Now a summation over δ accomplishes the proof of Theorem 3.1.

§4 Average value of $L\left(\frac{1}{2}, \chi_D\right)$

In this section, we outline the proof of Jutila [Jut2] of the mean and mean-square of $L(\frac{1}{2}, \chi_D)$. The results are as follows.

Theorem 4.1 *We have*

$$\sum_{0 < d \leq Y}^{*} L(\frac{1}{2}, \chi_d) = c_1 Y \log Y + c_2 Y + \mathbf{O}(Y^{3/4+\epsilon})$$

where the sum is over fundamental discriminants and c_1 is a non-zero constant.

Theorem 4.2 *We have*

$$\sideset{}{^*}\sum_{0 < d \leq Y} L(\tfrac{1}{2}, \chi_d)^2 = c_3 Y (\log Y)^3 + \mathbf{O}(Y(\log Y)^{5/2+\epsilon})$$

where the sum is over fundamental discriminants and c_3 is a non-zero constant.

Similar statements hold for negative discriminants also.

Corollary 4.3 *Let $N(Y)$ denote the number of fundamental discriminants $0 < d \leq Y$ such that $L(\frac{1}{2}, \chi_d) \neq 0$. Then*

$$N(Y) \gg Y/\log Y.$$

Proof. By Cauchy's inequality,

$$\left| \sideset{}{^*}\sum_{0 < d \leq Y} L(\tfrac{1}{2}, \chi_d) \right|^2 \ll \left(\sideset{}{^*}\sum_{0 < d \leq Y} L(\tfrac{1}{2}, \chi_d)^2 \right) \cdot N(Y).$$

By Theorems 4.1 and 4.2, this implies that

$$Y^2 (\log Y)^2 \ll Y(\log Y)^3 N(Y).$$

The result follows.

Proof of Theorem 4.1. Let

$$f_Y(n, w) = \sideset{}{^*}\sum_{0 < d \leq Y} \left(\frac{d}{n}\right) d^w$$

the summation ranging over fundamental discriminants, and

$$f_Y(n) = f_Y(n, 0).$$

For a real primitive character $\chi \bmod q > 1$, and an $X \geq 1$ we have the identity

$$L(\tfrac{1}{2}, \chi) = \sum_{n=1}^{\infty} \chi(n) \exp(-n/X) n^{-1/2}$$

$$- \frac{1}{2\pi i} \int_{(-\frac{1}{2}-\epsilon)} L(\tfrac{1}{2} - s, \chi) \left(\frac{q}{\pi}\right)^{-s} \frac{\Gamma(\frac{1}{2}(a + \frac{1}{2} - s))}{\Gamma(\frac{1}{2}(a + \frac{1}{2} + s))} \Gamma(s) X^s ds$$

which follows easily from the functional equation. Here, $a = \frac{1}{2}(1 - \chi(-1))$. Summing over χ corresponding to $d > 0$, and observing that $a = 0$ in this case, we get

$$\sideset{}{^*}\sum_{0 < d \leq Y} L(\tfrac{1}{2}, \chi_d) = \sum_{n=1}^{\infty} f_Y(n) \exp(-n/X) n^{-1/2}$$

$$- \frac{1}{2\pi i} \int_{(-\frac{1}{2}-\epsilon)} \sum_{n=1}^{\infty} \left(f_Y(n, -s) n^{s-\frac{1}{2}} \right) \pi^s \frac{\Gamma(\frac{1}{4} - \frac{s}{2})}{\Gamma(\frac{1}{4} + \frac{s}{2})} \Gamma(s) X^s ds$$

$$= S - I$$

say. When n is a square, we have

$$f_Y(n) = c_n Y + \mathbf{O}(Y^{\frac{1}{2}}\mathbf{d}(n))$$

where

$$c_n = \frac{3}{\pi^2} \prod_{p|n}(1 + \frac{1}{p})^{-1}.$$

Moreover, for $\mathrm{Re}(s) > 0$,

$$f_Y(n, s) = c_n \frac{Y^{1+s}}{1+s} + \mathbf{O}((|s| + 1)\mathbf{d}(n)Y^{\frac{1}{2}+\sigma}).$$

We apply this first to S. Indeed, in S the squares contribute an amount

$$\sum_{m=1}^{\infty}(c_m Y + \mathbf{O}(Y^{\frac{1}{2}}\mathbf{d}(m^2)))\exp(-m^2/X)m^{-1}$$

$$= Y\sum_{m=1}^{\infty}\frac{c_m}{m}\exp(-m^2/X) + \mathbf{O}(Y^{\frac{1}{2}}(\log X)^3).$$

From Exercise 5, we know that

$$\sum_{m=1}^{\infty}\frac{c_m}{m}\exp(-m^2/X) = \frac{3}{\pi^2}\prod_p\left(1 - \frac{1}{p(p+1)}\right)\left(\frac{1}{2}\log X + c\right) + \mathbf{O}(X^{-\frac{1}{2}+\epsilon})$$

for some constant c. For non-square values of n, we apply the estimate of Exercise 3, namely

$$\sideset{}{'}\sum_{1<n\leq N}\left|\sideset{}{^*}\sum_{0<d\leq Y}\left(\frac{d}{n}\right)\right|^2 \ll NY(\log N)^4. \tag{4.1}$$

Here the outer sum ranges over non-square integers and the inner sum over fundamental discriminants. By Cauchy's inequality, we get

$$\sideset{}{'}\sum_{n=1}^{\infty}f_Y(n)\exp(-n/X)n^{-1/2} \ll (\sideset{}{'}\sum_{n=1}^{\infty}f_Y(n)^2\exp(-n/X))^{\frac{1}{2}}(\log X)^{\frac{1}{2}}.$$

Now using (4.1) and partial summation, we deduce that the right hand side is

$$\ll Y^{1/2}X^{1/2}(\log X)^{5/2}.$$

Thus

$$S = c_1 Y \log X + c'Y + \mathbf{O}(Y^{\frac{1}{2}}X^{\frac{1}{2}}(\log X)^{5/2})$$

where

$$c_1 = \frac{3}{2\pi^2}\prod_p\left(1 - \frac{1}{p(p+1)}\right) \neq 0$$

and c' is some constant. In I, the squares contribute an amount

$$-\frac{1}{2\pi i} \int_{(-\frac{1}{2}-\epsilon)} \sum_{m=1}^{\infty} m^{2s-1} \left(c_m \frac{Y^{1-s}}{1-s} + \mathbf{O}((|s|+1)\mathrm{d}(m^2)Y^{1+\epsilon}) \right)$$

$$\pi^s \frac{\Gamma(\frac{1}{4} - \frac{s}{2})}{\Gamma(\frac{1}{4} + \frac{s}{2})} \Gamma(s) X^s ds.$$

Using the fact that

$$\sum_{m=1}^{\infty} c_m m^{-s} = \frac{3}{\pi^2} P(s)\zeta(s)$$

where

$$P(s) = \prod_p \left(1 - \frac{1}{(p+1)p^s} \right),$$

we see that the above integral gives

$$-\frac{1}{2\pi i} \int_{(-\frac{1}{2}-\epsilon)} \frac{Y^{1-s}}{1-s} \frac{3}{\pi^2} \zeta(1-2s) P(1-2s)\pi^s \frac{\Gamma(\frac{1}{4} - \frac{s}{2})}{\Gamma(\frac{1}{4} + \frac{s}{2})} \Gamma(s) X^s ds + \mathbf{O}(Y^{1+\epsilon} X^{-\frac{1}{2}-\epsilon}).$$

An easy calculation shows that

$$\zeta(1-2s)\Gamma(s) = -\frac{1}{2s^2} + \frac{3\gamma/2}{s} - \gamma^2 + \cdots$$

and so the main term above is

$$-Y\{c'' + \frac{3}{\pi^2} \frac{P(1)}{2} \log X/Y + \mathbf{O}((X/Y)^{\frac{1}{2}-\epsilon}\epsilon^{-1})\}$$

for some constant c''. Next, we observe that when summation is restricted to the non-squares, the integrand is

$$\ll \exp(-|t|)X^{-1/2-\epsilon} \sum_{\substack{n=1 \\ n\neq m^2}}^{\infty} {}'|f_Y(n, -s)|n^{-1-\epsilon}$$

where t is the imaginary part of s. Using the above character sum estimate (4.1) and partial summation, it follows that for $\sigma \geq 0$ we have

$$\sum_{\substack{n\leq N \\ n\neq m^2}} {}'|f_Y(n, s)|^2 \ll (|s|+1)^2 NY^{1+2\sigma}(\log N)^4.$$

Hence the integral over t is

$$\ll \epsilon^{-1} Y^{1+\epsilon} X^{-1/2-\epsilon}.$$

Hence

$$I = -Y\{c'' + \frac{3}{\pi^2} \frac{P(1)}{2} \log X/Y + \mathbf{O}((X/Y)^{\frac{1}{2}-\epsilon}\epsilon^{-1})\} + \mathbf{O}(\epsilon^{-1}Y^{1+\epsilon} X^{-\frac{1}{2}-\epsilon}).$$

Now choosing $X = Y^{\frac{1}{2}}$ gives the result. This proves Theorem 4.1.

For the proof of Theorem 4.2, we need several lemmas.

Lemma 4.4   *We have the following estimate.*

$$\sideset{}{'}\sum_{|D|\leq X}\left|\sum_{n\leq X}\mathbf{d}(n)\chi_0^{(k)}(n)\left(\frac{D}{n}\right)\right| \ll YX^{\frac{1}{2}}\mathbf{d}(k)^2(\log XY)^{17}.$$

Proof. See exercise 6.

Lemma 4.5   *For any $A > 0$, we have the estimate*

$$\sideset{}{'}\sum_{n\leq X}\mathbf{d}(n)\sideset{}{^*}\sum_{0<d\leq Y}\left(\frac{d}{n}\right) \ll Y^{\frac{1}{2}}X(\log XY)^{5/2}\log\log XY + YX^{\frac{1}{2}}(\log XY)^{-A}$$

where the sum is taken over integers n which are not squares and fundamental discriminants d.

Proof. Let us consider the double sum where the inner sum is taken over $d \equiv 1(\mathrm{mod}\,4)$. The other cases are similar. The sum in question is

$$\sideset{}{'}\sum_{n\leq X}\mathbf{d}(n)\sum_{\substack{m\leq Y\\m\equiv 1(\mathrm{mod}\,4)}}\left(\sum_{a^2|m}\mu(a)\right)\left(\frac{m}{n}\right)$$

$$=\sum_{a\leq Y^{\frac{1}{2}}}\mu(a)\chi_0^{(2)}(a)\sideset{}{'}\sum_{n\leq X}\mathbf{d}(n)\chi_0^{(a)}(n)\sum_{\substack{h\leq Y/a^2\\h\equiv 1(\mathrm{mod}\,4)}}\left(\frac{h}{n}\right)$$

$$= S_1 + S_2$$

where the sum S_1 is over $a \leq Z = (\log XY)^B$ for a suitable constant $B > 0$ and S_2 is the remainder. Applying Cauchy's inequality twice, we get

$$S_1 \leq \left(\sum_{a\leq Z}a^{-1}\right)^{\frac{1}{2}}\left(\sum_{a\leq Z}a\left|\sideset{}{'}\sum_{n\leq X}\mathbf{d}(n)\chi_0^{(a)}(n)\sum_{\substack{h\leq Ya^{-2}\\h\equiv 1(\mathrm{mod}\,4)}}\left(\frac{h}{n}\right)\right|^2\right)^{\frac{1}{2}}$$

$$\ll (\log Z)^{\frac{1}{2}}\left\{\sum_a a\left(\sum_n\mathbf{d}(n)^2\right)\left(\sideset{}{'}\sum_n\left|\sum_h\left(\frac{h}{n}\right)\right|^2\right)\right\}^{\frac{1}{2}}$$

$$\ll X^{\frac{1}{2}}(\log X)^{3/2}(\log\log XY)^{\frac{1}{2}}\left(\sum_a a\sideset{}{'}\sum_n\left|\sum_h\left(\frac{h}{n}\right)\right|^2\right)^{\frac{1}{2}}.$$

The double sum over h and n is estimated by (4.1) above. We find it is

$$\sideset{}{'}\sum_n\left|\sum_h\left(\frac{h}{n}\right)\right|^2 \ll YXa^{-2}(\log X)^2.$$

Hence, summing over a we get

$$S_1 \ll Y^{\frac{1}{2}} X (\log X)^{5/2} \log \log XY.$$

In S_2 we sum first over n using Lemma 4.4. Taking into account also the values of h which are squares, we get

$$S_2 \ll \sum_{Z < a \leq Y^{\frac{1}{2}}} (Ya^{-2} X^{\frac{1}{2}} \mathbf{d}(a)^2 (\log XY)^{17} + Y^{\frac{1}{2}} a^{-1} X (\log X))$$

$$\ll YX^{\frac{1}{2}} (\log XY)^{20-B} + Y^{\frac{1}{2}} X (\log XY)^2.$$

Combining the above estimates we get the stated result.

Now we are ready to establish the mean square of $L(\frac{1}{2}, \chi_d)$.

Proof of Theorem 4.2. Again from the functional equation, we have the following identity:

$$\sideset{}{^*}\sum_{0 < d \leq Y} L(\frac{1}{2}, \chi_d)^2$$

$$= \sum_{n=1}^{\infty} f_Y(n) \mathbf{d}(n) \exp(-n/X) n^{-\frac{1}{2}}$$

$$- \frac{1}{2\pi i} \int_{(-3/4)} \{\sum_{n=1}^{\infty} f_Y(n, -2s) \mathbf{d}(n) n^{s-\frac{1}{2}}\} \frac{\Gamma^2(\frac{1}{4} - \frac{s}{2})}{\Gamma^2(\frac{1}{4} + \frac{s}{2})} \Gamma(s) (\pi^2 X)^s ds$$

$$= S(X, Y) + I(X, Y).$$

Using Lemma 4.5, the non-squares in $S(X, Y)$ contribute an amount

$$\ll \sqrt{XY} (\log XY)^{5/2} \log \log XY + Y (\log XY)^{-A}.$$

For the integral, the sum over the non-square values of n is split at a point $U \geq Y$ say. For the initial segment, the line of integration is moved to $-\eta$ (say) for some $0 < \eta < \frac{1}{2}$. This gives an estimate

$$\ll U^{\frac{1}{2}} Y^{\frac{1}{2}} \left(\frac{Y^2}{UX}\right)^{\eta} (\log UY)^{5/2} (\log \log UY).$$

For the tail, one again uses Lemma 4.5 and partial summation to get an estimate

$$\ll U^{\frac{1}{2}} Y^{\frac{1}{2}} \left(\frac{Y^2}{UX}\right)^{3/4} (\log UY)^{5/2} \log \log UY.$$

If we choose $U = X = Y$, the error terms give a total contribution of

$$\ll Y (\log Y)^{5/2} (\log \log Y).$$

Finally, it is easy to check that the squares in the sum over n in $S(X,Y)$ and in $I(X,Y)$ give a total contribution of

$$cY (\log Y)^3 + \mathbf{O}(Y(\log Y)^2)$$

where

$$c = \frac{1}{8\pi^2} \prod_p \left(1 - \frac{4p^2 - 3p + 1}{p^4 + p^3} \right) \neq 0.$$

Combining all of these estimates proves Theorem 4.2.

§5 Non-vanishing for a positive proportion of characters, I

We shall discuss the following result from [BM].

Theorem 5.1 *Suppose that q is prime and sufficiently large. Unconditionally,*

$$\#\{\chi \,(\mathrm{mod}\, q) : L(1/2, \chi) \neq 0\} \geq c\phi(q)$$

where $c \geq .04$.

The assumption on q can be weakened significantly but only at the cost of making the proof more intricate due to the presence of imprimitive characters.

To prove the theorem, we have to consider a mollifier polynomial

$$M(s, \chi) = \sum_{n \leq Z} \lambda(n) \chi(n) n^{-s}$$

where $\lambda(n)$ are the Barban-Vehov weights and Z is a parameter to be chosen. More precisely, we will choose $Z = \sqrt{q}$, and $Y = (\log q)$ and

$$\lambda(n) = \begin{cases} \mu(n) & 1 \leq n \leq Y \\ \mu(n) \frac{\log Z/n}{\log Z/Y} & Y \leq n \leq Z \\ 0 & n \geq Z. \end{cases}$$

The mollifier polynomial is supposed to be a good approximation to $1/L(s, \chi)$. Let us also set

$$a(n) = \sum_{d|n} \lambda(d).$$

Observe that

$$a(1) = 1, \quad a(n) = 0 \ for \ 1 < n \leq Y.$$

Now, we consider the integral

$$\frac{1}{2\pi i}\int_{(2)} L(\frac{1}{2}+w,\chi)M(\frac{1}{2}+w,\chi)X^w\Gamma(w)dw.$$

Here X is a parameter to be specified later. We find that the integral is

$$S(\chi)=\sum_{n=1}^{\infty}\frac{a(n)\chi(n)}{n^{\frac{1}{2}}}\exp(-\frac{n}{X}).$$

On the other hand, moving back the line of integration, we see that it is

$$L(\frac{1}{2},\chi)M(\frac{1}{2},\chi)+\frac{1}{2\pi i}\int_{(-\eta)} L(\frac{1}{2}+w,\chi)M(\frac{1}{2}+w,\chi)X^w\Gamma(w)dw.$$

Applying the functional equation in the integral, we get

$$\frac{1}{2\pi i}\int_{(-\eta)} L(\frac{1}{2}-w,\bar\chi)M(\frac{1}{2}+w,\chi)\gamma(\frac{1}{2}+w,\chi)X^w\Gamma(w)dw$$

where

$$\gamma(s,\chi)=\frac{g(\chi)}{i^a q^{\frac{1}{2}}}\left(\frac{2}{\pi}\right)^{\frac{1}{2}}\left(\frac{2\pi}{q}\right)^{s-\frac{1}{2}}\sin\left(\frac{\pi}{2}(a+s)\right)\Gamma(1-s)$$

where $a=0,1$ and $\chi(-1)=(-1)^a$ and $g(\chi)$ is the Gauss sum determined by χ. In applying the functional equation, we are assuming that χ is primitive. Now, we can expand $L(\frac{1}{2}-w,\bar\chi)$ as a Dirichlet series provided $\eta>\frac{1}{2}$. We split this series at Z and so we get two integrals and the formula

$$S(\chi)=L(\frac{1}{2},\chi)M(\frac{1}{2},\chi)+I(\chi)+J(\chi).$$

Now we compare mean and mean square estimates for $S(\chi),I(\chi)$ and $J(\chi)$ and show that they are incompatible with the frequent vanishing of $L(\frac{1}{2},\chi)$. Choosing $X=q$, we prove the following estimates:

$$\sum S(\chi)\sim\phi(q) \tag{1}$$

$$\sum|S(\chi)|^2<\frac{5}{2}\phi(q) \tag{2}$$

$$\sum|I(\chi)|^2\sim c\phi(q) \tag{3}$$

where

$$c=\frac{4}{\pi^2}\prod_{p>2}\left(1+\frac{2}{(p-2)(p+1)}\right)\prod_{p>2}\left(1-\frac{1}{(p-1)^2}\right)\sim.374750$$

and

$$\sum|J(\chi)|\ll\frac{q}{\log q}. \tag{4}$$

What is the role of the specific mollifier weights in all of the above estimates? The key points are that by an important result of S. Graham [Gra], we have

$$\sum_{n \leq N} |a(n)|^2 \sim \begin{cases} \frac{N \log N/Y}{\log^2 Z/Y} & Y < N < Z \\ \frac{N}{\log Z/Y} & Z \leq N \end{cases}$$

and one uses this as well as a partial generalization of it to arithmetic progressions. A second key point is that if we sum instead

$$\sum a(n)a(n-k)$$

we save an extra logarithm over the above estimates.

Using (1)–(4), let us see how to deduce Theorem 5.1. We have

$$\sum S(\chi) = \sum_{\chi : L(\frac{1}{2},\chi)=0} (I(\chi) + J(\chi)) + \sum_{\chi : L(\frac{1}{2},\chi) \neq 0} S(\chi) \sim \phi(q).$$

Now,

$$| \sum_{\chi : L(\frac{1}{2},\chi)=0} I(\chi) + J(\chi)| \leq \sum |I(\chi)| + \sum |J(\chi)|$$

$$\leq \phi(q)^{1/2} \left(\sum |I(\chi)|^2 \right)^{1/2} + \mathbf{O}(\frac{q}{\log q})$$

$$\lesssim \sqrt{c}\phi(q).$$

Therefore,

$$\sum_{\chi : L(\frac{1}{2},\chi) \neq 0} S(\chi) \gtrsim (1 - \sqrt{c})\phi(q).$$

On the other hand, by Cauchy-Schwarz,

$$\sum_{\chi : L(\frac{1}{2},\chi) \neq 0} S(\chi) \leq \#\{\chi : L(\frac{1}{2},\chi) \neq 0\}^{1/2} \left(\sum |S(\chi)|^2 \right)^{1/2}.$$

Thus,

$$\#\{\chi : L(\frac{1}{2},\chi) \neq 0\} \gtrsim \frac{2}{5}(1 - \sqrt{c})^2\phi(q).$$

Now we shall give a few more details on this outline. We begin with two results on the Barban-Vehov weights .

Let $1 \leq z_1 \leq z_2$. Following Barban and Vehov [BV], we introduce the functions

$$\Lambda_i(n) = \begin{cases} \mu(n) \log(z_i/n) & \text{if } n \leq z_i \\ 0 & \text{if } n > z_i, \end{cases}$$

for $i = 1, 2$. We also define

$$\lambda(n) = \frac{\Lambda_2(n) - \Lambda_1(n)}{\log(z_2/z_1)} = \begin{cases} \mu(n) & 1 \leq n \leq z_1 \\ \mu(n) \frac{\log(z_2/n)}{\log(z_2/z_1)} & z_1 \leq n \leq z_2 \\ 0 & n > z_2. \end{cases} \qquad (5.1)$$

Let us define

$$a(n) = \sum_{d \mid n} \lambda(d).$$

Graham [Gra] has found asymptotic estimates for the mean square of the $a(n)$. We recall his main result.

Proposition 5.2   *We have*

$$\sum_{n \leq N} |a(n)|^2 = \begin{cases} \frac{N \log(N/z_1)}{\log^2(z_2/z_1)} + \mathbf{O}\left(\frac{N}{\log^2(z_2/z_1)}\right) & \text{if } z_1 < N < z_2 \\ \frac{N}{\log(z_2/z_1)} + \mathbf{O}\left(\frac{N}{\log^2(z_2/z_1)}\right) & \text{if } z_2 \leq N. \end{cases}$$

Applying the Cauchy-Schwarz inequality and Proposition 5.2, we deduce the following.

Proposition 5.3   *Let $r \leq N$ be prime and $(b, r) = 1$. We have*

$$\sum_{\substack{b < n \leq N \\ n \equiv b \pmod{r}}} |a(n)| \ll \frac{N}{\phi(r)^{1/2}(\log z_2/z_1)^{1/2}}.$$

We next obtain an estimate for a shifted convolution.

Proposition 5.4   *Let $1 \leq k \in \mathbb{Z}$, $t \in \mathbb{R}$ and $k \leq M < N$. Then we have*

$$\sum_{M < n \leq N} a(n)a(n-k) \ll \frac{k}{\phi(k)} \frac{N + z_2^2}{(\log z_2/z_1)^2}.$$

The proof will require two preliminary results. We begin by recalling a result from Graham [Gra, Lemma 2].

Lemma 5.5   *For any integer r, and any $c > 0$,*

$$\sum_{\substack{n \leq Q \\ (n,r)=1}} \frac{\mu(n)}{n} \log\left(\frac{Q}{n}\right) = \frac{r}{\phi(r)} + \mathbf{O}_c\left(\sigma_{-\frac{1}{2}}(r)\log^{-c}(2Q)\right).$$

Lemma 5.6   *We have for $1 \leq d_1, d_2 \leq z_2$ and $r_1, r_2 \geq 1$ that*

$$\sum_{\substack{1 \leq j_1 \leq z_1/d_1, \ 1 \leq j_2 \leq z_2/d_2 \\ (j_1,j_2)=(j_1,r_1)=(j_2,r_2)=1}} \frac{\Lambda_1(d_1 j_1)\Lambda_2(d_2 j_2)}{j_1 j_2}$$

$$\ll \left(\frac{d_1 r_1}{\phi(d_1 r_1)} + \sigma_{-\frac{1}{2}}(d_1 r_1)\right)\left(\frac{d_2 r_2}{\phi(d_2 r_2)} + \sigma_{-\frac{1}{2}}(d_2 r_2)\right).$$

The same estimate holds even if we drop the condition that $(j_1, j_2) = 1$.

Proof. The sum in question is

$$\sum \frac{\Lambda_1(d_1 j_1)\Lambda_2(d_2 j_2)}{j_1 j_2} \sum_{e|(j_1,j_2)} \mu(e) = \sum_{e \leq z_1/d_1} \mu(e) \sum \frac{\Lambda_1(d_1 j_1)\Lambda_2(d_2 j_2)}{j_1 j_2},$$

the inner sum ranging over j_1, j_2 satisfying

$$1 \leq j_1 \leq z_1/d_1, \qquad 1 \leq j_2 \leq z_2/d_2$$
$$j_1, j_2 \equiv 0 \ (\mathrm{mod}\, e), (j_1, r_1) = (j_2, r_2) = 1.$$

Let us set $r = r_1 r_2$ and $d = d_1 d_2$. Then the sum is seen to be

$$\sum_{\substack{e \leq z_1/d_1 \\ (e,r)=1}} \frac{\mu(e)}{e^2} \sum_{\substack{\ell_1 \leq z_1/d_1 e, \ \ell_2 \leq z_2/d_2 e \\ (\ell_1,r_1)=(\ell_2,r_2)=1}} \frac{\Lambda_1(d_1 e \ell_1)\Lambda_2(d_2 e \ell_2)}{\ell_1 \ell_2}$$

$$= \mu(d_1)\mu(d_2) \sum_{\substack{e \leq z_1/d_1 \\ (e,dr)=1}} \frac{\mu(e)}{e^2} \left\{ \sum_{\substack{\ell_1 \leq z_1/d_1 e \\ (\ell_1,d_1 e r_1)=1}} \frac{\mu(\ell_1)\log(z_1/d_1 e \ell_1)}{\ell_1} \right\}$$

$$\times \left\{ \sum_{\substack{\ell_2 \leq z_2/d_2 e \\ (\ell_2,d_2 e r_2)=1}} \frac{\mu(\ell_2)\log(z_2/d_2 e \ell_2)}{\ell_2} \right\}$$

$$= \mu(d_1)\mu(d_2) \sum_{\substack{e \leq z_1/d_1 \\ (e,dr)=1}} \frac{\mu(e)}{e^2} \prod_{k=1}^{2} \left\{ \frac{d_k e r_k}{\phi(d_k e r_k)} + \mathbf{O}_c\left(\sigma_{-\frac{1}{2}}(d_k e r_k) \log^{-c}\left(\frac{2z_k}{d_k e}\right) \right) \right\}$$

using Lemma 5.5.

The main terms contribute an amount

$$\frac{\mu(d_1)\mu(d_2)dr}{\phi(d_1 r_1)\phi(d_2 r_2)} \cdot \sum_{\substack{e \leq z_1/d_1 \\ (e,dr)=1}} \frac{\mu(e)}{\phi(e)^2} \ll \frac{dr}{\phi(d_1 r_1)\phi(d_2 r_2)}.$$

The product of the **O**-terms contributes an amount

$$\ll \sum \frac{1}{e^2} \sigma_{-\frac{1}{2}}(d_1 r_1)\sigma_{\frac{1}{2}}(d_2 r_2)\sigma_{-\frac{1}{2}}(e)^2 \ll \sigma_{-\frac{1}{2}}(d_1 r_1)\sigma_{\frac{1}{2}}(d_2 r_2).$$

The cross-terms contribute an amount

$$\ll \sum_{\substack{e \leq z_1/d_1 \\ (e,dr)=1}} \frac{1}{e^2}\left\{ \frac{d_1 e r_1}{\phi(d_1 e r_1)} \cdot \sigma_{-\frac{1}{2}}(d_2 e r_2) \log^{-c}\left(\frac{2z_2}{d_2 e}\right) \right.$$

$$\left. + \frac{d_2 e r_2}{\phi(d_2 e r_2)} \cdot \sigma_{-\frac{1}{2}}(d_2 e r_2) \log^{-c}\left(\frac{2z_2}{d_2 e}\right) \right\}$$

$$\ll \left\{ \frac{d_1 r_1}{\phi(d_1 r_1)}\sigma_{-\frac{1}{2}}(d_2 r_2) + \frac{d_2 r_2}{\phi(d_2 r_2)}\sigma_{-\frac{1}{2}}(d_1 r_1) \right\}$$

since the series $\sum \sigma_{-\frac{1}{2}}(e)/e\phi(e)$ converges. This proves the first statement. The second statement is easy to verify since there is now no condition relating j_1 and j_2. We argue as above setting $e = 1$.

Now we are ready to prove the estimate of the shifted convolution.

Proof of Proposition 5.4. Again, we consider the sum

$$\sum_{M < n \leq N} \left(\sum_{d|n} \Lambda_1(d) \right) \left(\sum_{e|n-k} \Lambda_2(e) \right) \tag{5.2}$$

and we find that it is equal to

$$\sum_{d,e} \Lambda_1(d)\Lambda_2(e) \sum_{\substack{M < n \leq N \\ n \equiv 0 \pmod d \\ n \equiv k \pmod e}} 1. \tag{5.3}$$

We see that the inner sum is zero unless $(d,e)|k$ and if (d,e) divides k it is

$$\frac{N - M}{[d,e]} + \mathbf{O}(1).$$

Thus, (5.3) is

$$(N - M) \sum_{\substack{d,e \\ (d,e)|k}} \frac{\Lambda_1(d)\Lambda_2(e)}{[d,e]} + +\mathbf{O}\Big(\sum_{\substack{d,e \\ (d,e)|k}} |\Lambda_1(d)\Lambda_2(e)| \Big) \tag{5.4}$$

The **O**-term is easily seen to be

$$\ll z_1 z_2.$$

To evaluate the main term, we see that the sum over d, e is

$$\sum_{\substack{d,e \\ (d,e)|k}} \frac{\Lambda_1(d)\Lambda_2(e)}{de} \sum_{\substack{m|d \\ m|e}} \phi(m). \tag{5.5}$$

This is seen to be equal to

$$\sum_{m|k} \frac{\phi(m)}{m^2} \sum_{d_0,e_0} \frac{\Lambda_1(md_0)\Lambda_2(me_0)}{d_0 e_0}.$$

Here, the inner sum ranges over pairs d_0, e_0 satisfying

$$1 \leq d_0 \leq \frac{z_1}{m}, \qquad 1 \leq e_0 \leq \frac{z_2}{m} \qquad (d_0, m) = (e_0, m) = 1.$$

Also note that in the outer sum m must be squarefree for otherwise $\Lambda_1(md_0) = \Lambda_2(me_0) = 0$. Thus, invoking Lemma 5.6, we find that the main term in (5.5) is

$$\ll \sum_{m|k} \frac{\mu^2(m)\phi(m)}{m^2} \left(\frac{m}{\phi(m)} + \sigma_{-\frac{1}{2}}(m) \right)^2$$

$$\ll \frac{k}{\phi(k)}.$$

Hence the main term in (5.4) is

$$\ll \frac{k}{\phi(k)}N.$$

Summarizing, (5.3) is

$$\ll \frac{k}{\phi(k)}N + z_1 z_2.$$

The proposition follows.

Next, we discuss the mollifier polynomial. We introduce the parameters

$$Y = (\log q)$$
$$Z = q^{1/2}$$

Corresponding to the choices $z_1 = Y$ and $z_2 = Z$, we have the weights

$$\lambda(n) = \frac{\Lambda_2(n) - \Lambda_1(n)}{\log(Z/Y)}.$$

We define the Dirichlet polynomial

$$M(s,\chi) = \sum_{n \leq Z} \frac{\lambda(n)\chi(n)}{n^s}$$

where χ is a Dirichlet character. Then, we have

$$L(s,\chi)M(s,\chi) = \sum_{n=1}^{\infty} \frac{a(n)\chi(n)}{n^s}$$

where

$$a(n) = \sum_{d|n} \lambda(d)$$

satisfies

$$a(1) = 1$$
$$a(n) = 0 \quad \text{for} \quad 1 < n \leq Y.$$

We record the following estimate.

Lemma 5.7 *For $\sigma \neq \frac{1}{2}$, we have*

$$\sum_{\chi (\mathrm{mod}\, q)} |M(s,\chi)|^2 \ll \frac{(q+Z)}{(1-2\sigma)} \cdot (q^{\frac{1}{2}-\sigma} \cdot \frac{1}{(\log q)^2} + Y^{1-2\sigma}).$$

Proof. We use the large sieve inequality [D] to get

$$\sum_{\chi (\mathrm{mod}\, q)} |M(s,\chi)|^2 \ll (Z+q) \sum_{n \leq Z} \frac{|\lambda(n)|^2}{n^{2\sigma}}$$

$$\ll (q+Z) \left\{ \sum_{n \leq Y} \frac{1}{n^{2\sigma}} + \sum_{Y < n \leq Z} \left(\frac{\log Z/n}{\log Z/Y} \right)^2 \cdot \frac{1}{n^{2\sigma}} \right\}$$

$$\ll (q+Z) \left\{ \frac{Y^{1-2\sigma}}{1-2\sigma} + \frac{Z^{1-2\sigma}}{1-2\sigma} \cdot \frac{1}{(\log Z/Y)^2} \right\}.$$

The result follows from our choices of Y and Z.

Now we consider the basic equation that relates the above quantities. Let us define

$$S(s,\chi) = S(s,\chi,q) = \sum_{n=1}^{\infty} \frac{a(n)\chi(n)}{n^s} e^{-n/q}.$$

Let $s \in \mathbb{C}$ with $1 > \sigma = \mathrm{Re}(s) \geq \frac{1}{2}$. Using the well-known identity

$$\frac{1}{2\pi i} \int_{(2)} X^w \Gamma(w)\, dw = e^{-1/X},$$

we find that for a character χ,

$$S(s,\chi) = \frac{1}{2\pi i} \int_{(2)} L(s+w,\chi) M(s+w,\chi) q^w \Gamma(w)\, dw.$$

Moving the line of integration to the left, we find that

$$S(s,\chi) = L(s,\chi) M(s,\chi) + \frac{1}{2\pi i} \int_{(-\eta)} L(s+w,\chi) M(s+w,\chi) q^w \Gamma(w)\, dw \quad (5.6)$$

where $\sigma < \eta < 1$.

We can decompose the integral along the line $-\eta$ into two parts as follows. Suppose that χ is non-trivial. We apply the functional equation

$$L(s,\chi) = \gamma(s,\chi) L(1-s,\bar{\chi})$$

where
$$\gamma(s,\chi) = \frac{g(\chi)}{i^a q^{\frac{1}{2}}} \left(\frac{2}{\pi}\right)^{\frac{1}{2}} \left(\frac{2\pi}{q}\right)^{s-\frac{1}{2}} \sin\left(\frac{\pi}{2}(a+s)\right) \Gamma(1-s).$$

(Here $g(\chi)$ is the Gauss sum, $a = 0,1$ and $\chi(-1) = (-1)^a$.) Then we truncate the Dirichlet series expansion of $L(1-s-w, \overline{\chi})$ at Z. Let us set

$$I(s,\chi) = \frac{1}{2\pi i} \int_{(-\eta)} \gamma(s+w,\chi) \left\{ \sum_{n<Z} \frac{\overline{\chi}(n)}{n^{1-s-w}} \right\} M(s+w,\chi) q^w \Gamma(w) dw$$

and

$$J(s,\chi) = \frac{1}{2\pi i} \int_{(-\eta)} \gamma(s+w,\chi) \left\{ \sum_{n\geq Z} \frac{\overline{\chi}(n)}{n^{1-s-w}} \right\} M(s+w,\chi) q^w \Gamma(w) dw$$

Thus, we get
$$S(s,\chi) = L(s,\chi)M(s,\chi) + I(s,\chi) + J(s,\chi). \tag{5.7}$$

If $L(s,\chi) = 0$, then $S(s,\chi)$ and $I(s,\chi) + J(s,\chi)$ are equal. We will therefore try to show that, in general, they are *not* equal and for this purpose we study their mean values. We begin with $J(s,\chi)$ which is the easiest of the three to estimate.

Proposition 5.8   *For $|\operatorname{Im} s| < 1$ and $0 \leq \sigma \leq 1$, we have*

$$\sum_{1\neq\chi(\bmod q)} |J(s,\chi)| \ll_\epsilon \frac{q^{\frac{3}{2}-\sigma}}{\log q}$$

Proof. From Stirling's formula, we know that

$$\gamma(s,\chi) \ll (q(|s|+1))^{\frac{1}{2}-\sigma}.$$

Using this and the definition, we find that

$$\sum_{1\neq\chi(\bmod q)} |J(s,\chi)| \ll q^{\frac{1}{2}-\sigma+\eta} q^{-\eta} \sum_{\chi(\bmod q)} \int_{(-\eta)} (|w|+1)^{\frac{1}{2}-\sigma+\eta}$$

$$|\sum_{n\geq Z} \frac{\overline{\chi}(n)}{n^{1-s-w}}| \|M(s+w,\chi)\| |\Gamma(w)| |dw|$$

which by a double application of the Cauchy-Schwarz inequality is

$$\ll q^{\frac{1}{2}-\sigma} \sum_{\chi(\bmod q)} \left(\int (|w|+1)^{1-2\sigma+2\eta} |\sum_{n\geq Z} \frac{\overline{\chi}(n)}{n^{1-s-w}}|^2 |\Gamma(w)| |dw| \right)^{\frac{1}{2}}$$

$$\times \left(\int |M(s+w,\chi)|^2 |\Gamma(w)| |dw| \right)^{\frac{1}{2}}$$

$$\ll q^{\frac{1}{2}-\sigma}\left(\sum_{\chi(\bmod q)}\int(|w|+1)^{1-2\sigma+2\eta}|\sum_{n\geq Z}\frac{\overline{\chi}(n)}{n^{1-s-w}}|^2|\Gamma(w)||dw|\right)^{\frac{1}{2}}\times$$

$$\left(\sum_{\chi(\bmod q)}\int|M(s+w,\chi)|^2|\Gamma(w)||dw|\right)^{\frac{1}{2}}.$$

Using the large sieve inequality and Lemma 5.7, we find that

$$\sum_{1\neq\chi(\bmod q)}|J(s,\chi)|\ll q^{\frac{1}{2}-\sigma}\left\{\sum_{n\geq Z}(q+n)n^{2(\sigma-\eta-1)}\right\}^{\frac{1}{2}}\times$$

$$\times\left\{\frac{(q+Z)}{1-2(\sigma-\eta)}\cdot(\frac{q^{\frac{1}{2}-\sigma+\eta}}{(\log q)^2}+Y^{1-2(\sigma-\eta)})\right\}^{\frac{1}{2}}$$

$$\ll q^{\frac{1}{2}-\sigma}Z^{\sigma-\eta}\left\{\frac{q}{Z}\frac{1}{|2(\sigma-\eta)-1|}+\frac{1}{|\sigma-\eta|}\right\}^{\frac{1}{2}}\times$$

$$\times\left\{\frac{(q+Z)^{\frac{1}{2}}}{|2(\sigma-\eta)-1|^{\frac{1}{2}}}\frac{q^{\frac{1}{2}(\frac{1}{2}-\sigma+\eta)}}{\log q}\right\}.$$

Now, let us choose η so that it satisfies

$$\frac{1}{4}>|\eta-\sigma|>\frac{1}{8}\quad\text{(say)}$$

if $\sigma<\frac{3}{4}$.

We would then have

$$\sum_{1\neq\chi(\bmod q)}|J(s,\chi)|\ll\frac{q^{\frac{3}{2}-\sigma}}{\log q}\qquad(5.8)$$

which proves the result.

§6 Non-vanishing for a positive proportion, II

Next, we study the mean and mean square of $S(s,\chi)$

Proposition 6.1 *For any $\epsilon>0$, we have*

$$\sum_{\chi\,(\bmod q)}S(s,\chi)=\phi(q)+\mathbf{O}_\epsilon(q^{1-\sigma+\epsilon}).$$

Moreover, the same estimate holds if we sum only over non-trivial characters.

Proof. By definition, we have that

$$\sum_{\chi \pmod q} S(s,\chi) = \sum_{n=1}^{\infty} \frac{a(n)}{n^s} e^{-n/q} \sum_{\chi \pmod q} \chi(n)$$

$$= \phi(q) \sum_{\substack{n=1 \\ n \equiv 1 \pmod q}}^{\infty} \frac{a(n)}{n^s} e^{-n/q}.$$

Using the bound $|a(n)| \le d(n) \ll_\epsilon n^\epsilon$, we find that the sum is

$$e^{-1/q} + \mathbf{O}_\epsilon \left(\frac{1}{q^{\sigma-\epsilon}} \sum_{t=1}^{\infty} t^{\epsilon-\sigma} \exp(-t) \right).$$

The **O**-term is

$$\ll_\epsilon q^{-\sigma+\epsilon}.$$

It thus follows that

$$\sum_{\chi \pmod q} S(s,\chi) = \phi(q) + \mathbf{O}_\epsilon(q^{1-\sigma+\epsilon}).$$

Finally,

$$S(s,1) = \sum_{(n,q)=1} \frac{a(n)}{n^s} e^{-n/q} \ll q^{1-\sigma+\epsilon}$$

as before. This proves the result.

Proposition 6.2   *We have*

$$\sum_{\chi \pmod q} |S(\tfrac{1}{2},\chi)|^2 = \frac{5}{2}\phi(q) + \mathbf{O}(q(\log q)^{-\frac{1}{2}}).$$

Proof. We see that the sum is equal to

$$\sum_{n_1,n_2=1}^{\infty} \frac{a(n_1)a(n_2)}{(n_1 n_2)^{\frac{1}{2}}} \exp(-(n_1+n_2)/q) \sum_{\chi \pmod q} \chi(n_1)\overline{\chi}(n_2)$$

which is seen to be

$$\phi(q) \sum_{n_1,n_2=1}^{\infty}{}' \frac{a(n_1)a(n_2)}{(n_1 n_2)^{\frac{1}{2}}} \exp(-(n_1+n_2)/q), \qquad (6.1)$$

where the sum ranges over pairs (n_1, n_2) satisfying

$$n_1 \equiv n_2 (\text{mod } q), \qquad (n_1, q) = (n_2, q) = 1$$

We split the double sum into three pieces $\Sigma_1 + \Sigma_2 + \Sigma_3$. In Σ_1, we have $n_1 < n_2$, in Σ_2 we have $n_1 > n_2$, and in Σ_3 we have $n_1 = n_2$. The estimation of Σ_1 and Σ_2 is the same, so we only consider Σ_1. We have

$$\Sigma_1 = \sum_{\substack{n_1=1 \\ (n_1,q)=1}}^{\infty} \frac{a(n_1) \exp(-n_1/q)}{n_1^{\frac{1}{2}}} \sum_{\substack{n_2=1 \\ n_2 \equiv n_1 (\text{mod } q) \\ n_2 > n_1}}^{\infty} \frac{a(n_2) \exp(-n_2/q)}{n_2^{\frac{1}{2}}}. \tag{6.2}$$

We begin by considering the sum over n_2. We must necessarily have $n_2 > q$ for if $n_2 \leq q$, then $n_1 \leq q$ also and so the congruence $n_2 \equiv n_1 (\text{mod } q)$ would force $n_1 = n_2$. We split Σ_1 into three subsums Σ_{11}, Σ_{12} and Σ_{13} where

in Σ_{11} we have $n_2 \geq q \log q$

in Σ_{12} we have $q \leq n_1 < q \log q$ and $n_1 < n_2 < q \log q$.

in Σ_{13} we have $n_1 < q$ and $q < n_2 < q \log q$.

In Σ_{11}, we see, by partial summation, that the sum over n_2 is

$$\ll q^{-1} \int_{q \log q}^{\infty} \left\{ \sum_{\substack{n_1 < n \leq u \\ n \equiv n_1 (\text{mod } q)}} |a(n)| \right\} u^{-\frac{1}{2}} e^{-u/q} du.$$

We have from Proposition 5.3 that

$$\sum_{\substack{n_1 < n \leq u \\ n \equiv n_1 (\text{mod } q)}} |a(n)| \ll \frac{u}{\phi(q)^{\frac{1}{2}} (\log q)^{\frac{1}{2}}}.$$

Thus, we find that the integral is

$$\ll \frac{1}{q^{3/2} (\log q)^{1/2}} \int_{q \log q}^{\infty} u^{\frac{1}{2}} e^{-u/q} du$$

and this is

$$\ll (\log q)^{-1/2} \int_{\log q}^{\infty} v^{\frac{1}{2}} e^{-v} dv$$

$$\ll q^{-1}.$$

Inserting this into the n_1-sum, using Proposition 5.2, the Cauchy-Schwarz inequality and partial summation, we have

$$\Sigma_{11} \ll \frac{q^{\frac{1}{2}}}{(\log q)^{\frac{1}{2}}} \frac{1}{q}$$

$$\ll q^{-\frac{1}{2}} (\log q)^{-\frac{1}{2}}.$$

Now we consider the contribution of Σ_{12}. This is

$$\sum_{\substack{q \leq n_1 < q \log q \\ (n_1,q)=1}} \frac{a(n_1)e^{-n_1/q}}{n_1^{\frac{1}{2}}} \sum_{\substack{n_1 < n_2 < q \log q \\ n_2 \equiv n_1 (\mathrm{mod}\, q)}} \frac{a(n_2)e^{-n_2/q}}{n_2^{\frac{1}{2}}}.$$

We split the n_1 sum into $\mathbf{O}(\log \log q)$ sums of the form

$$\sum_{\substack{U < n_1 \leq 2U \\ (n_1,q)=1}} \frac{a(n_1)e^{-n_1/q}}{n_1^{\frac{1}{2}}} \sum_{\substack{n_1 < n_2 < q \log q \\ n_2 \equiv n_1 (\mathrm{mod}\, q)}} \frac{a(n_2)e^{-n_2/q}}{n_2^{\frac{1}{2}}}.$$

Let us write $n_2 = n_1 + jq$. The above double sum may therefore be written as

$$\sum_{j < \log q} e^{-j} \sum_{\substack{U < n_1 \leq 2U \\ (n_1,q)=1}} e^{-2n_1/q} \frac{a(n_1)a(n_1 + jq)}{n_1^{\frac{1}{2}}(n_1 + jq)^{\frac{1}{2}}} \qquad (6.3)$$

If we drop the condition $(n_1, q) = 1$, then we introduce an additional sum

$$\sum_{j < \log q} e^{-j} \sum_{U < qk \leq 2U} e^{-2k} \frac{a(kq)a((k+j)q)}{(kq)^{\frac{1}{2}}((k+j)q)^{\frac{1}{2}}}. \qquad (6.4)$$

Observe that as q is prime, and $\lambda(n) = 0$ for $n > Z = q^{1/2}$, we have

$$a(kq) = \sum_{d|kq} \lambda(d) = \sum_{d|k} \lambda(d) = a(k).$$

Therefore, we have the estimate

$$|a(kq)| \leq \mathbf{d}(k) \ll_\epsilon k^\epsilon.$$

A similar estimate holds for $a((k+j)q)$. Using this in (6.4), we see that it is

$$\ll q^{-1} \sum_{j < \log q} e^{-j} \sum_{U < qk \leq 2U} \frac{e^{-2k}}{k^{\frac{1}{2}-\epsilon}(k+j)^{\frac{1}{2}-\epsilon}}$$

and this is

$$\ll q^{-1}.$$

The sum in (6.3) may thus be replaced by

$$\sum_{j < \log q} e^{-j} \sum_{U < n_1 \leq 2U} e^{-2n_1/q} \frac{a(n_1)a(n_1 + jq)}{n_1^{\frac{1}{2}}(n_1 + jq)^{\frac{1}{2}}} \qquad (6.5)$$

Let us set

$$G(u) = \sum_{U < n_1 \leq u} a(n_1)a(n_1 + jq).$$

By Proposition 5.4, we see that for $U < u$,

$$G(u) \ll \frac{j}{\phi(j)} \left(\frac{(u + (j+1)q)}{(\log q)^2} \right).$$

The sum over n_1 in (6.5) can be estimated using partial summation. We find that it is equal to

$$\left. \frac{G(u)e^{-\frac{2u}{q}}}{u^{\frac{1}{2}}(u+jq)^{\frac{1}{2}}} \right|_U^{2U} + \int_U^{2U} G(u)d\left(\frac{e^{-\frac{2u}{q}}}{u^{\frac{1}{2}}(u+jq)^{\frac{1}{2}}} \right).$$

Using the estimate for $G(u)$ quoted above, we see that this is

$$e^{-2U/q} \frac{j}{\phi(j)} \frac{(U+jq)^{\frac{1}{2}}}{U^{\frac{1}{2}}} \frac{U}{q} (\log q)^{-2}.$$

Incorporating these estimates into the sum over j, we find that (6.5) is for $\sigma \neq 1$

$$\ll \sum_{j < \log q} \frac{(U+jq)^{\frac{1}{2}}U^{\frac{1}{2}}}{q} e^{-2U/q}(\log q)^{-2} \frac{j}{\phi(j)} e^{-j}$$

which is

$$\ll \frac{U^{\frac{1}{2}}e^{-2U/q}}{q(\log q)^2} \sum_{j < \log q} e^{-j} \frac{j}{\phi(j)} (U+jq)^{\frac{1}{2}}$$

$$\ll q^{\frac{1}{2}}(\log q)^{-3/2} \frac{U^{\frac{1}{2}}}{q} e^{-2U/q}.$$

Now summing this over U, we find it is

$$\ll (\log q)^{-3/2}.$$

Now we discuss the contribution of Σ_{13}. By the Cauchy-Schwarz inequality, we see that

$$|\Sigma_{13}|^2 \ll \left(\sum_{n_1 < q} \frac{a(n_1)^2}{n_1} \exp(-2n_1/q) \right) \left(\sum_{n_1 \leq q} \left| \sum_{\substack{q < n_2 < q \log q \\ n_2 \equiv n_1 (\bmod q)}} \frac{a(n_2)e^{-n_2/q}}{n_2^{\frac{1}{2}}} \right|^2 \right).$$

The first factor above is $\mathbf{O}(1)$ as can be seen from our discussion of Σ_3 below. As for the second factor, we see that it is equal to

$$\sum_{\substack{q < n_2, n_2' < q \log q \\ n_2 \equiv n_2' (\mathrm{mod}\, q)}} \frac{a(n_2) e^{-n_2/q}}{n_2^{\frac{1}{2}}} \frac{a(n_2') e^{-n_2'/q}}{(n_2')^{\frac{1}{2}}}.$$

Again, we split this sum into three sums according as $n_2 < n_2'$, $n_2 = n_2'$, and $n_2 > n_2'$. The third is the same as the first. Also, we note that the first sum is just Σ_{12} which we have estimated above as being

$$\ll (\log q)^{-3/2}.$$

As for the second, we see that it is equal to

$$\sum_{q \leq n_2 < q \log q} \frac{a(n_2)^2 e^{-2n_2/q}}{n_2}.$$

Using Proposition 5.2 and partial summation, this is

$$\ll (\log q)^{-1}.$$

Inserting this into the above, we deduce that

$$\Sigma_{13} \ll (\log q)^{-3/4}.$$

Finally, we discuss the estimation of Σ_3, namely the terms with $n_1 = n_2$. Thus,

$$\Sigma_3 = \sum_{\substack{n=1 \\ (n,q)=1}}^{\infty} \frac{a(n)^2}{n} \exp(-2n/q) = \sum_{n \leq Y} + \sum_{Y < n \leq q} + \sum_{n > q}. \tag{6.6}$$

Since $a(n) = 0$ for $1 < n \leq Y$, we have

$$\sum_{n \leq Y} = \exp(-2/q) \tag{6.7}$$

Also, by partial summation and Proposition 5.2, we find that

$$\sum_{n > q} \ll (\log(z_2/z_1))^{-1}. \tag{6.8}$$

Thus, we see from (6.6) - (6.8) that

$$\Sigma_3 = 1 + \sum_{\substack{Y < n \leq q \\ (n,q)=1}} \frac{a(n)^2}{n} \exp(-2n/q) + \mathbf{O}((\log q)^{-1})$$

Let us denote the sum on the right by S. We find that

$$S = \sum_{Y < n \leq q} \frac{a(n)^2}{n} \left(1 + \mathbf{O}\left(\frac{n}{q}\right)\right).$$

Now, the **O**-term is

$$\ll \frac{1}{q} \sum_{\substack{Y < n \leq q \\ (n,q)=1}} a(n)^2$$

$$\ll \frac{1}{q} \frac{1}{\log(z_2/z_1)} q$$

$$\ll (\log q)^{-1}.$$

The main term is equal to

$$\sum_{Y < n < q} \frac{a(n)^2}{n}.$$

Finally, using Proposition 5.2, we get

$$\sum_{n < q} \frac{a(n)^2}{n} = \sum_{1 \leq n \leq Y} + \sum_{Y < n \leq Z} + \sum_{Z < n < q} \frac{a(n)^2}{n}.$$

The first sum is equal to 1 since $a(n) = 0$ for $1 < n \leq Y$. Using Proposition 5.2 and partial summation, we see that the second sum is

$$\sum_{Y < n \leq Z} \frac{(\log n/Y)}{(\log Z/Y)^2} \cdot \frac{1}{n} + \mathbf{O}\left(\frac{1}{\log Z/Y}\right).$$

$$= \frac{1}{2} + \mathbf{O}\left(\frac{1}{\log q}\right).$$

Similarly, the third sum is

$$\sum_{Z < n < q} \frac{1}{\log Z/Y} \frac{1}{n} + \mathbf{O}\left(\frac{1}{\log q}\right)$$

which is

$$= 1 + \mathbf{O}\left(\frac{1}{\log q}\right).$$

Putting these together we deduce that

$$\sum_{n < q} \frac{a(n)^2}{n} = \frac{5}{2}\left(1 + \mathbf{O}(\frac{1}{\log q})\right).$$

This completes the proof of the proposition.

We need also a result on the mean square of $I(\frac{1}{2}, \chi)$. We only state the result and refer the reader to [BM, Proposition 10.1] for the proof.

Proposition 6.3   *We have*

$$\sum |I(\tfrac{1}{2},\chi)|^2 = c\phi(q) + \mathbf{O}(\frac{q\log\log q}{\log q}).$$

Now we can put together the above results to prove Theorem 5.1.

Proof of Theorem 5.1. Let $s_0 \in \mathbb{C}$ satisfy $\frac{1}{2} \le \operatorname{Re} s_0 < 1 - \frac{1}{\log q}$. We return now to (5.7). For $\chi \ne 1$, we have

$$S(s_0,\chi) = L(s_0,\chi)M(s_0,\chi) + I(s_0,\chi) + J(s_0,\chi).$$

Thus,

$$\sum_{\chi \ne 1} S(s_0,\chi) = {\sum}'\big(I(s_0,\chi) + J(s_0,\chi)\big) + {\sum}'' S(s_0,\chi)$$

where $\sum'$ ranges over $\chi \ne 1$ such that $L(s_0,\chi) = 0$ and $\sum''$ over the remaining non-trivial $\chi(\operatorname{mod} q)$. By Proposition 6.1, we have

$$\sum_{\chi \ne 1} S(s_0,\chi) = \phi(q) + \mathbf{O}_\epsilon\big(q^{1-\sigma+\epsilon}\big).$$

Thus, we have

$${\sum}'' S(s_0,\chi) = \phi(q) - {\sum}'\big(I(s_0,\chi) + J(s_0,\chi)\big) + \mathbf{O}_\epsilon\big(q^{1-\sigma_0+\epsilon}\big)$$

and consequently,

$$\begin{aligned}{\sum}'' S(s_0,\chi) \ge {}& \phi(q) - |{\sum}'\big(I(s_0,\chi) + J(s_0,\chi)\big)| \\ & + \mathbf{O}_\epsilon\big(q^{1-\sigma+\epsilon}\big)\end{aligned}$$

Now, assuming $|\operatorname{Im} s_0| < 1$,

$$\begin{aligned}|{\sum}'\big(I(s_0,\chi) + J(s_0,\chi)\big)| \le {}& \sum |I(s_0,\chi)| + \sum |J(s_0,\chi)| \\ \le {}& \phi(q)^{\frac{1}{2}}\big(\sum |I(s_0,\chi)|^2\big)^{\frac{1}{2}} \\ & + \mathbf{O}\left(\frac{q^{\frac{3}{2}-\sigma}}{\log q}\right)\end{aligned}$$

by Proposition 5.8. Now using Proposition 6.3, we have

$$|{\sum}'(I(\tfrac{1}{2},\chi) + J(\tfrac{1}{2},\chi))| \le \sqrt{c}\phi(q) + \mathbf{O}\left(q\left(\frac{\log\log q}{\log q}\right)^{\frac{1}{2}}\right) + \mathbf{O}(q(\log q)^{-1}).$$

Thus,

$$|\sum{}''S(\tfrac{1}{2},\chi)| \geq (1-\sqrt{c})\phi(q) + \mathbf{O}(q^{\frac{1}{2}+\epsilon}) + \mathbf{O}\left(q\left(\frac{\log\log q}{\log q}\right)^{\frac{1}{2}}\right)$$
$$+ \mathbf{O}(q(\log q)^{-1}).$$

On the other hand, by the Cauchy-Schwarz inequality, setting $\mathcal{N}(s_0, q)$ to be the number of $\chi(\bmod q)$ with $L(s_0, \chi) \neq 0$, we get

$$|\sum{}''S(s_0,\chi)|^2 \leq \mathcal{N}(s_0,q)\left(\sum|S(s_0,\chi)|^2\right).$$

We have from Proposition 6.2

$$\sum_{\chi(\bmod q)} |S(\tfrac{1}{2},\chi)|^2 = \frac{5}{2}\phi(q) + \mathbf{O}(q(\log q)^{-\frac{1}{2}}).$$

We deduce that

$$\frac{2}{5}\phi(q)(1-\sqrt{c})^2 + \mathbf{O}\left(q\sqrt{\frac{\log\log q}{\log q}}\right) \leq \mathcal{N}(\tfrac{1}{2},q)\left(1 + \mathbf{O}((\log q)^{-\frac{1}{2}})\right).$$

Thus,

$$\mathcal{N}(\tfrac{1}{2},q) \geq \frac{2}{5}\phi(q)(1-\sqrt{c})^2 + \mathbf{O}\left(q\sqrt{\frac{\log\log q}{\log q}}\right).$$

The methods of this section in fact prove a significantly stronger result [BM, Theorem 11.1]. We state it below.

Theorem 6.4 *Fix a σ in the interval $\frac{1}{2} \leq \sigma < 1$. Then, for all sufficiently large primes q,*

$$L(\sigma, \chi) \neq 0$$

for a positive proportion of the characters $\chi(\bmod q)$.

Remark. The proof produces a lower bound for this proportion. How large q must be taken will depend on σ.

Finally, we state another result [BM, Theorem 12.1] which can be proved by refining the techniques described above.

Theorem 6.5 *Let q be a sufficiently large prime. There is an absolute constant $c > 0$ such that for a positive proportion of the characters $\chi \bmod q$, $L(\sigma, \chi) \neq 0$ in the interval*

$$\frac{1}{2} + \frac{c}{\log q} \leq \sigma \leq 1.$$

§7 A conditional improvement

It is an interesting question to ask whether Theorem 5.1 can be strengthened by assuming the Riemann Hypothesis. Of course, the Riemann Hypothesis tells us that there are no zeroes for $\mathrm{Re}(s) > \frac{1}{2}$, but it gives no direct information about $s = \frac{1}{2}$. The following result is due to R. Murty [RM].

Theorem 7.1   *Let q be a prime. Assume the Riemann Hypothesis for all the Dirichlet L-functions $L(s,\chi)$. The number of characters $\chi \bmod q$ with $L(\frac{1}{2},\chi) \neq 0$ is at least $\frac{1}{2}\phi(q)$.*

The proof depends on the explicit formula method. Let us write

$$-\frac{L'}{L}(s,\chi) = \sum_{n=1}^{\infty} \Lambda(n)\chi(n)n^{-s}.$$

Proposition 7.2   *Let F be a function satisfying the following hypotheses:*
(i)   For some $\epsilon > 0$, $F(x)\exp((1+\epsilon)x)$ is integrable and of bounded variation.
(ii)   The function $(F(x) - F(0))/x$ is of bounded variation.
For such a function define the transform

$$\phi(\gamma) = \int_{-\infty}^{\infty} F(x)e^{i\gamma x}dx.$$

Then

$$\sum_{\gamma} \phi(\gamma) = 2F(0)\log\frac{\sqrt{q}}{2\pi} + \phi(-i/2)\delta_\chi - \frac{1}{\pi}\int_{-\infty}^{\infty}\frac{\Gamma'}{\Gamma}(1+it)\phi(t)dt$$

$$- 2\sum_{n=1}^{\infty}\frac{\Lambda(n)\chi(n)}{n}F(\log n)$$

where the sum on the left hand side is over γ such that $L(\frac{1}{2}+i\gamma,\chi) = 0$ and $\frac{1}{2} \leq \mathrm{Re}(\frac{1}{2}+i\gamma) \leq 1$ and

$$\delta_\chi = \begin{cases} 1 & \text{if } \chi \text{ is the principal character} \\ 0 & \text{otherwise.} \end{cases}$$

The following two lemmas are proved easily by straightforward calculations.

Lemma 7.3   *Let $T > 0$ and define*

$$F(x) = \begin{cases} 2T - |x| & \text{if } |x| \leq 2T \\ 0 & \text{otherwise.} \end{cases}$$

Then F satisfies the conditions of Proposition 7.2 and

$$\phi(\gamma) = \left(\frac{2\sin(\gamma T)}{\gamma}\right)^2.$$

Lemma 7.4 *Let $T > 1$. Then*

$$\int_0^\infty \frac{\Gamma'}{\Gamma}(1+it)\left(\frac{\sin tT}{t}\right)^2 dt \ll T.$$

Proof of Theorem 7.1. We choose the function F given in Lemma 7.3 and apply the explicit formula to the Dedekind zeta function of $\mathbb{Q}(\zeta_q)$:

$$Z(s) = \prod_{\chi \,(\mathrm{mod}\, q)} L(s, \chi).$$

As we are assuming the Riemann Hypothesis, all of the γ are real. Let us set

$$T = \frac{1}{2}\log x.$$

Let r_χ denote the order of zero of $L(s, \chi)$ at the point $s = \frac{1}{2}$. Then we have the inequality

$$\sum_\chi r_\chi (\log x)^2 \le 2\phi(q)(\log x)(\log \sqrt{q}/2\pi) + 4x^{\frac{1}{2}} + \mathbf{O}(\phi(q)\log x) - 2\phi(q)$$

$$\times \sum_{\substack{n \le x \\ n \equiv 1 (\mathrm{mod}\, q)}} \frac{1}{n}\Lambda(n)\log\frac{x}{n}.$$

We can discard the last sum on the right as it is non-negative. Now choosing

$$x = \phi(q)^2$$

gives

$$\sum_\chi r_\chi \le \frac{1}{2}\phi(q) + \mathbf{O}(\phi(q)/\log q).$$

This proves the result.

Exercises

1.   Prove the Polya-Vinogradov estimate for imprimitive characters as follows. Let $\chi \bmod q$ be induced from $\chi_1 \bmod q_1$ and write $q = q_1 r$. Then

$$\sum_{n \le x} \chi(n) = \sum_{\substack{n \le x \\ (n,r)=1}} \chi_1(n).$$

Express the condition $(n, r) = 1$ in terms of the Möbius function to deduce that the quantity to be estimated is

$$\sum_{d|r} \mu(d)\chi_1(d) \sum_{m \le x/d} \chi_1(m).$$

Now apply the Polya-Vinogradov estimate for the inner sum and deduce that the above quantity is

$$\ll q_1^{\frac{1}{2}} \log q_1 \sum_{d|r} |\mu(d)| \ll q_1^{\frac{1}{2}} \log q_1.$$

2. Prove the estimate of Theorem 3.1 in the case $Y \leq X^{\frac{1}{2}}$ as follows.
    (i)   First,

$$S(X,Y) \leq \sum_{\substack{r,s \leq Y \\ rs \neq a^2}} {\sum_{|D| \leq X}}' \left(\frac{D}{rs}\right) + \sum_{n^2 \leq Y^2} \mathbf{d}(n^2) {\sum_{|D| \leq X}}' 1.$$

Show that

$$\sum_{n \leq Y} \mathbf{d}(n^2) \ll Y (\log Y)^2$$

to deduce that the second sum is $\ll XY(\log Y)^2$.
    (ii) As for the first sum, prove that if we include square values of D in the inner sum, we increase the whole quantity only by an amount which is $\ll Y^2 X^{\frac{1}{2}} \leq XY$.
    (iii) View the inner sum as

$${\sum_{|D| \leq X}}^* \chi(D)$$

where $\chi(D) = (D/rs)$ is a character of conductor $\leq 4rs$ and the sum is over integers D in the specified range and which satisfy (ii) in the definition of $\mathcal{D}$. Show that this sum can now be expressed by a finite number of character sums with nonprincipal characters of conductor $\leq 16rs$. Apply the Polya-Vinogradov estimate to deduce that the sum over D is $\ll (rs)^{\frac{1}{2}} \log Y$. Now sum over r, s to deduce that the whole sum is $\ll Y^3 \log X \leq XY \log X$.

3. Prove the following variant of the estimate of Theorem 3.1:

$${\sum_{|D| \leq X}}' \left| {\sum_{n \leq Y}}^* \left(\frac{D}{n}\right) \right|^2 \ll XY(\log X)^4.$$

Here, the sum over n is restricted to fundamental discriminants. To deduce this from Theorem 3.1, begin by writing the inner sum as

$${\sum_{n \leq Y}}^* \left(\frac{D}{n}\right) = \sum_{n \leq Y} \left(\sum_{d^2|n} \mu(d) \right) \left(\frac{D}{n}\right)$$

and rearrange as

$$\sum_{d \leq Y^{\frac{1}{2}}} \mu(d) \sum_{\substack{n \leq Y \\ d^2 \mid n}} \left(\frac{D}{n}\right).$$

4. Prove that when n is a square,

$$\sideset{}{'}\sum_{0 < d \leq Y} \left(\frac{d}{n}\right) d^s = c_n \frac{Y^{1+s}}{1+s} + \mathbf{O}((|s|+1)\mathbf{d}(n)Y^{\frac{1}{2}+\sigma})$$

for $\sigma > 0$. Here, the summation is over fundamental discriminants.

5. Prove that for some constant c, we have

$$\sum_{m=1}^{\infty} \frac{c_m}{m} \exp(-m^2/X) = \frac{3}{\pi^2} \prod_p \left(1 - \frac{1}{p(p+1)}\right) \left(\frac{1}{2}\log X + c\right) + \mathbf{O}(X^{-\frac{1}{2}+\epsilon}).$$

6. Using the mean-value estimate of Jutila [Jut3]

$$\sideset{}{'}\sum_{|D| \leq X} \int_{-T}^{T} |L(\tfrac{1}{2}+it, \chi)|^2 \ll XT(\log XT)^{16}$$

deduce the estimate of Lemma 4.4.

References

[B]   R. Balasubramanian, A note on Dirichlet's L-functions *Acta Arith.*, **38** (1980), 273–283

[BM] R. Balasubramanian and V. Kumar Murty, Zeros of Dirichlet L-functions, *Ann. Scient. Ecole Norm. Sup.*, **25** (1992), 567–615

[BV] M.B. Barban and P.P. Vehov, On an extremal problem *Trans. Moscow Math. Soc.*, **18** (1968), 91–99

[D]   H. Davenport, Multiplicative Number Theory, Springer-Verlag, 1980.

[Gra] S. Graham, An asymptotic estimate related to Selberg's sieve *J. Numb. Thy.*, **10** (1978), 83–94

[HB] R. Heath-Brown, The fourth power mean of Dirichlet's L-functions, *Analysis*, **1** (1981), 25–32

[Hil] A. Hildebrand, Large values of character sums, *J. Numb. Thy.*, **29** (1988), 271–296

[Jut1] M. Jutila, On character sums and class numbers, *J. Numb. Thy.*, **5** (1973), 203–214

[Jut2] M. Jutila, On the mean value of $L(\frac{1}{2}, \chi)$ for real characters, *Analysis*, **1** (1981), 149–161

[Jut3] M. Jutila, On mean values of L-functions and short character sums with real characters, *Acta Arith.*, **26** (1975), 405–410

[KM] V. Kumar Murty, Non-vanishing of L-functions and their derivatives, in: Automorphic forms and analytic number theory, pp. 89–113, ed. R. Murty, CRM, Montreal, 1990.

[Me] J.-F. Mestre, Formules explicites et minorations de conducteurs de variétés algébriques, *Comp. Math.*, **58** (1986), 209–232

[MV] H.L. Montgomery and R.C. Vaughan, Exponential sums with multiplicative coefficients, *Invent. Math.*, **43** (1977), 69–82

[RM] M. Ram Murty, Simple zeroes of L-functions, in: Number Theory, ed. R. Mollin, pp. 427–439, de Gruyter, 1989.

[Sie] C. L. Siegel, On the zeros of the Dirichlet L-functions, *Annals of Math.*, **46** (1945), 409–422

[Vino] I. M. Vinogradov, The method of trigonometrical sums in the theory of numbers, Interscience, London-New York, 1955.

Chapter 6
Non-Vanishing of Quadratic Twists
of Modular L-Functions

§1 Introduction

Statement of results

Let f be a holomorphic cusp form for $\Gamma_0(N)$ of weight 2 and character ϵ. We assume that f is a normalized newform for the Hecke operators. Denote by $L(s,f)$ the L-function attached to f. For $\mathrm{Re}(s) > 3/2$, it is given by an absolutely convergent Dirichlet series

$$L(s,f) = \sum_{n=1}^{\infty} \frac{a(n)}{n^s}.$$

For $L(s,f)$ we have the functional equation

$$A^s \Gamma(s) L(s,f) = \omega A^{2-s} \Gamma(2-s) L(2-s,\bar{f}),$$

where $A = \sqrt{N}/2\pi$, ω is a complex number of absolute value 1 and $L(s,\bar{f})$ is defined by

$$L(s,\bar{f}) = \sum_{n=1}^{\infty} \frac{\bar{a}(n)}{n^s}$$

for $\mathrm{Re}(s) > 3/2$ and by analytic continuation for all values of s. (This series actually converges (conditionally) for $\mathrm{Re}(s) > 5/6$. Thus, the functional equation and the Dirichlet series serve to define $L(s,f)$ for *all* values of s.)

For any D, let χ_D denote the quadratic character $\left(\frac{D}{\cdot}\right) \bmod D$. Let us set

$$L_D(s,f) = \sum_{n=1}^{\infty} \frac{a(n)\chi_D(n)}{n^s}.$$

Let D be a fundamental discriminant (that is, the discriminant of a quadratic field). Thus, $D \equiv 1 \pmod 4$ and D is squarefree, or $D = 4D_0$, $D_0 \equiv 2,3 \pmod 4$

and D_0 is squarefree. If $(D, N) = 1$, then $L_D(s, f)$ satisfies the functional equation

$$(A|D|)^s \Gamma(s) L_D(s, f) = \omega \chi_D(-N) \epsilon(D) (A|D|)^{2-s} \Gamma(2 - s) L_D(2 - s, \bar{f}).$$

The problem considered in this chapter is whether there exists a fundamental discriminant D prime to N such that $L_D(1, f) \neq 0$. This question has been considered in the case $\epsilon = 1$ and settled affirmatively by Waldspurger [W1], [W2]. Using a completely different approach, K. Murty [M] showed that the answer is *always* affirmative.

Theorem 1.1   *There exist infinitely many fundamental discriminants D such that*

$$L_D(1, f) \neq 0.$$

The purpose of this chapter is to give an exposition of this result. For forms with nontrivial character, Shimura [Sh] showed some years ago that there exists a twist (not necessarily quadratic) such that the twisted L-function does not vanish at $s = 1$. This has been generalized to number fields and general points by Rohrlich [Ro] who shows that given any point $s_0 \in \mathbf{C}$ there exists a twist (not necessarily quadratic) such that the twisted L-function does not vanish at s_0. It should be pointed out that in the context of number fields, there are examples of Waldspurger to show that in general, we may not be able to find a quadratic twist such that the twisted L-function does not vanish at the central critical point. Indeed, Waldspurger works with cuspidal automorphic representations of $\mathrm{PGL}(2)$ over any number field, and produces a necessary and sufficient condition for the existence of such a twist. This condition is always satisfied when the field is $\mathbb{Q}$ and the representation corresponds to a holomorphic modular form. Some references to more recent work on this question are given at the end of this section as well as in Chapter 8.

Though we have stated the result for forms of weight 2, we remark that it remains valid for holomorphic cusp forms of any weight ≥ 2.

We shall in fact prove the following result from which Theorem 1.1 will follow immediately.

Theorem 1.2   *Let $a \equiv 1 \pmod 4$, $(a, 2N) = 1$. Then,*

$$\frac{1}{Y} \int_1^Y \sum_{\substack{|D| \leq t \\ D \equiv a \pmod{8N}}} L_D(1, f) \, dt = C(f)Y + \mathbf{O}\left(\frac{Y}{(\log Y)^\nu}\right)$$

where $C(f) \neq 0$ and $0 < \nu < \rho = 1 - \frac{1}{5\sqrt{2}}(\sqrt{2} + 3\sqrt{3}) = .0652\ldots$.

Note that in Theorem 1.2, the sum is over *all* D with $|D| \leq t$, $D \equiv a \pmod{8N}$ and not only over fundamental discriminants. To deduce Theorem 1.1 from Theorem 1.2, one has only to check that if there were only a finite number of fundamental discriminants D_0 such that $L_{D_0}(1, f) \neq 0$, then the quantity estimated in Theorem 1.2 can be shown to be $\mathbf{O}(\sqrt{Y}(\log Y))$. (The interested reader is referred to Exercise 1 below, or to [MM, pp. 450–451] where a similar calculation is carried out in detail.)

Notation

We begin by introducing some notation. If D_0 is a fundamental discriminant which is coprime to N, then we may write the functional equation in the unsymmetric form

$$L_{D_0}(1+s, f) = \omega \chi_{D_0}(-N)\epsilon(D_0)|D_0|^{-2s} A^{-2s} \frac{\Gamma(1-s)}{\Gamma(1+s)} L_{D_0}(1-s, \bar{f}).$$

Let us define $\tilde{\mu}$ by

$$\frac{1}{L(s, f)} = \sum_{n=1}^{\infty} \frac{\tilde{\mu}(n, f)}{n^s}.$$

Any $D \equiv 1(\mathrm{mod}\, 4)$ can be written as, $D = D_0 \delta^2$, where D_0 is a fundamental discriminant and

$$L_D(s, f) = L_{D_0}(s, f) \sum_{d|\delta^2} \frac{\tilde{\mu}(d, f)}{d^s} \left(\frac{D_0}{d} \right).$$

For $\beta = \pm 1$, let us set

$$\tilde{f}_Y^{\beta}(n, s; a) = \sum_{\substack{0 < \beta D \le Y \\ D \equiv a(\mathrm{mod}\, 8N)}} \left(\frac{D}{n} \right) |D|^s, \quad D \text{ unrestricted}$$

$$f_Y^{\beta}(n, s; a) = \sum_{\substack{0 < \beta D_0 \le Y \\ D_0 \equiv a(\mathrm{mod}\, 8N)}} \left(\frac{D_0}{n} \right) |D_0|^s, \quad D_0 \text{ fundamental}$$

$$\tilde{f}_Y^{\beta}(n; a) = \tilde{f}_Y^{\beta}(n, 0; a), \qquad f_Y^{\beta}(n; a) = f_Y^{\beta}(n, 0; a).$$

Let us also set

$$\tilde{g}_Y^{\beta}(n, s; a) = \frac{1}{Y} \int_1^Y \tilde{f}_t^{\beta}(n, s; a)\, dt$$

and

$$g_Y^{\beta}(n, s; a) = \frac{1}{Y} \int_1^Y f_t(n, s; a) dt.$$

We shall write $\mathbf{d}(n)$ for the number of positive divisors of n. Also, b stands for a generic integer. Thus the statement $nd \ne b^2$ means that nd is not the square of an integer. For an integer n we shall often write $n = n_1 n_2$ where $(n_2, 2N) = 1$ and $p|n_1 \implies p|2N$. The complex numbers s_0 and s_1 have real parts σ_0 and σ_1 (respectively). We set

$$\log^+ x = \begin{cases} \log x & \text{if } x > 1 \\ 1 & \text{otherwise.} \end{cases}$$

Outline of Proof. Let D be an integer with $D \equiv 1 \pmod 4$. As in [MM], we begin by considering the integral

$$\frac{1}{2\pi i} \int_{(2)} L_D(1+s,f) X^s \Gamma(s) ds.$$

This is equal to

$$\sum_{m=1}^{\infty} \frac{a(m)}{m} \left(\frac{D}{m}\right) \exp(-m/X).$$

On the other hand, moving the line of integration to the left, we obtain a residue at $s=0$ equal to

$$L_D(1,f)$$

and an integral

$$\frac{1}{2\pi i} \int_{(-\eta)} L_D(1+s,f) X^s \Gamma(s) ds.$$

Here, $0 < \eta < 1$ and the integration is on the line $\text{Re}(s) = -\eta$.

In the above notation,

$$L_D(s,f) = L_{D_0}(s,f) \sum_{d \mid \delta^2} \frac{\tilde{\mu}(d,f)}{d^s} \left(\frac{D_0}{d}\right).$$

Applying the functional equation for $L_{D_0}(s,f)$ we see that the integral is

$$\epsilon(D_0) \omega \chi_{D_0}(-N) \times$$

$$\times \frac{1}{2\pi i} \int_{(-\eta)} |D_0|^{-2s} L_{D_0}(1-s,\bar{f}) \sum_{d \mid \delta^2} \frac{\tilde{\mu}(d,f)}{d^{1+s}} \left(\frac{D_0}{d}\right) \frac{\Gamma(1-s)}{\Gamma(1+s)} \left(\frac{X}{A^2}\right)^s \Gamma(s) ds.$$

Now, let us fix an integer a such that $a \equiv 1 \pmod 4$ and $(a,2N) = 1$. Unlike [MM], we now sum the above equation over all D satisfying $|D| \leq Y$ with $D \equiv a \pmod{8N}$. Thus, we include both positive and negative values of D. This is an important ingredient in controlling the error terms. (See the discussion later in this section and also Lemma 2.2 and Lemma 5.4). If we take $\eta > \frac{1}{2}$, then $L_{D_0}(1-s,\bar{f})$ is given by an absolutely convergent Dirichlet series. Inserting this and rearranging, we find that

$$\sum_{\substack{|D| \leq Y \\ D \equiv a \pmod{8N}}} L_D(1,f) = S(X,Y) + I(X,Y)$$

where

$$S(X,Y) = \sum_{m=1}^{\infty} \frac{a(m)}{m} \left\{ \tilde{f}_Y^+(m;a) + \tilde{f}_Y^-(m;a) \right\} \exp(-m/X)$$

and

$$I(X,Y)=\omega\epsilon(a)\left(\frac{a}{N}\right)\sum_{\substack{\delta\leq\sqrt{Y}\\(\delta,2N)=1}}\sum_{d|\delta^2}\bar{\epsilon}(\delta^2)\frac{\tilde{\mu}(d,f)}{d}\times$$

$$\times\frac{1}{2\pi i}\int_{(-\eta)}\sum\frac{\bar{a}(n)}{n^{1-s}}\left\{f^+_{Y/\delta^2}(nd,2s;a)-f^-_{Y/\delta^2}(nd,2s;a).\right\}\frac{\Gamma(1-s)}{\Gamma(1+s)}\left(\frac{X}{A^2d}\right)^s\Gamma(s)ds.$$

We shall prove that

$$\frac{1}{Y}\int_1^Y S(X,t)dt = C(f)Y + \mathbf{O}(X(\log X)^{-\rho})$$

where

$$C(f) = \frac{1}{8N}\sum_{n_1,n_2}\frac{a(n_1n_2^2)}{n_1n_2^2}\left(\frac{a}{n_1}\right)\frac{\phi(n_2)}{n_2}.$$

This is similar to the constant that occurs in [MM, Theorem 1] and an analogous argument shows that it is nonzero. Actually, we work with a more general series. For $(h, 2Nj) = 1$, and $\mathrm{Re}(s_0)$, $\mathrm{Re}(s_1) \geq 0$, define

$$C(s_0, s_1, j, h) = \frac{1}{8N(1 + 2s_0)(1 + s_0)}\sum_{\substack{n=n_1n_2\\n_2h=b^2\\(n_2h,j)=1}}\frac{a(n)}{n^{1+s_1}}\left(\frac{a}{n_1}\right)\frac{\phi(n_2h)}{n_2h}.$$

That the series converges in this domain follows from estimates in §3. Moreover, we note that

$$C(f) = C(0, 0, 1, 1).$$

The error estimate above is obtained by applying the integrated Polya-Vino-gradov inequality and an estimate of Rankin. For the integral, we show that a smoothened version of it is

$$\ll X^{\frac{1}{2}}Y^{\frac{1}{2}}\left(\frac{Y}{X}\right)^{4/5}(\log Y)^{\nu}$$

for any $0 < \nu < \rho/10$. Choosing $X = Y(\log Y)^{10\nu}$ in the above estimates, the main theorem follows.

The above estimate for the mean value of $I(X,t)$ is the technical heart of this chapter. It requires an integrated and refined version of the Polya-Vinogradov estimate (§2, §4), together with an iterated argument to estimate certain weighted sums of Fourier coefficients and Dirichlet characters (§6). Preliminary estimates for such weighted sums are obtained in §5. The treatment of the main terms is discussed in §3 and the theorems are proved in §7.

There are several estimates which we shall repeatedly use. Firstly, for Fourier coefficients, we have the estimate of Rankin

$$\sum_{n \leq X} |a(n)|^2 \ll X^2.$$

There is also the estimate of Rankin-Shahidi [Ra] (see Theorem IV.9.1)

$$\sum_{n \leq X} |a(n)| \ll X^{3/2} (\log X)^{-\rho}.$$

Secondly, for character sums, there is the estimate of Fainleib and Saparnijazov [FS] (see also [MM, Lemma 1]) which is a generalization of Theorem V.3.1 of Jutila [J]:

$$\sum_{\substack{n \leq X \\ n_2 \neq b^2}} \left| \sum_{\substack{0 < \beta h \leq Y \\ h \equiv a (\mathrm{mod}\, 8N)}} \left(\frac{h}{nd} \right) \right|^2 \ll (N^2/\phi(N)) d X Y (\log X d)^2.$$

Remarks on the proof. To prove Theorem 1.2, we consider averages over both positive and negative discriminants, and in addition, we introduce a further smoothing factor. Thus, we consider

$$\frac{1}{Y} \int_1^Y \sum_{\substack{|D| \leq t \\ D \equiv a (\mathrm{mod}\, 8N)}} L_D(1, f) \, dt = \sum_{\substack{|D| \leq t \\ D \equiv a (\mathrm{mod}\, 8N)}} \left(1 - \frac{|D|}{Y} \right) L_D(1, f).$$

If $\epsilon = 1$, then $\omega = \pm 1$ and the functional equation gives the relation

$$(1 - \omega \chi_D(-N)) L_D(1, f) = 0.$$

Thus, the requirement that $L_D(1, f) \neq 0$, imposes a condition on D. In particular, if $D \equiv a (\mathrm{mod}\, 8N)$ is fixed, we require sgn D to be chosen so that sgn $D = \omega \left(\frac{a}{N} \right)$.

In our arguments, the importance of choosing the correct value of sgn D is seen as follows. When we sum over D of a fixed sign, we expect (and get) contributions to the main term from (the analogues of) both the sum $S(X, Y)$ and the integral $I(X, Y)$. Together, this contribution is

$$\left(1 + \omega \left(\frac{a}{N} \right) (\mathrm{sgn}\, D) \right) cY$$

for a nonzero constant c. Thus, the "wrong" choice of sgn D would cause the main term to be cancelled. For essentially, the same reasons, summing over D of both signs doubles the contribution of $S(X, Y)$ to the main term and cancels the contribution of $I(X, Y)$. (See the proof of Theorem 1.2 in §7). If $\epsilon \neq 1$, then the root number does not give an obstruction to the choice of the sign of D. By including both values of sgn D, therefore, we get a statement valid in all cases.

There is a second, and deeper, reason for including both positive and negative values of D. In order to handle the error terms that come from $S(X, Y)$, we need to estimate character sums of the form

$$\sum_{\substack{|D| \leq Y \\ D \equiv a \,(\mathrm{mod}\, 8N)}} \left(\frac{D}{m}\right)$$

when m_2 is not a square. If χ is a nontrivial Dirichlet character modulo q, the Polya-Vinogradov estimate (Theorem V.2.1) is

$$\sum_{0 < \beta D \leq Y} \chi(D) \ll \sqrt{q}(\log q)$$

for $\beta = \pm 1$. If in addition, χ is even, Hua has shown (see Lemma 2.1 below) that

$$\frac{1}{Y} \int_1^Y \sum_{0 < \beta D \leq t} \chi(D) dt \ll \sqrt{q}.$$

Thus, for even characters, the extra averaging allows us to save a factor of $\log q$. Now, by summing over both positive and negative values of D, and integrating over t, we are able to filter out odd characters (see Lemma 2.2) and use this improved estimate.

To estimate the integral

$$\frac{1}{Y} \int_1^Y I(X, t) dt$$

we find that the device which filtered out odd characters in the sum has exactly the opposite effect in the integral. (This is because of the sgn character that is introduced by the functional equation). Therefore, we need an analogue of Hua's estimate which is valid for all characters. In §4, we give such an estimate in mean square. Indeed, a special case of Lemma 4.6 gives

$$\sum_{n \leq X} \left| \frac{1}{Y} \int_0^Y \sum_{\substack{0 < \beta D \leq t \\ D \equiv a \,(\mathrm{mod}\, 8N)}} \left(\frac{D}{n}\right) dt \right|^2 \ll X^2 (\log^+ \frac{X}{Y})^2$$

where

$$\log^+ a = \begin{cases} \log a & \text{if } a > 1 \\ 1 & \text{otherwise} \end{cases}$$

and $\beta = \pm 1$.

In addition to this, we need a sharp estimate for the quantity

$$\sum_{\substack{n \leq X \\ n_2 h \neq b^2}} \frac{a(n)}{n^{1+s_1}} \frac{1}{Y} \int_1^Y \left(f_t^+(nh, 2s_0; a) - f_t^-(nh, 2s_0; a) \right) \, dt.$$

It turns out that the required estimate is intimately connected with estimates for two other quantities. We shall now briefly describe these.

For the purpose of this exposition, we state them approximately as follows. (The reader who wishes a more precise statement is referred to the relevant sections.) Let $\alpha > 0$, $\beta = \pm 1$. We consider the following statements:

$$A^\beta(\alpha): \quad \frac{1}{Y} \int_1^Y \sum_{\substack{0 < \beta D \leq t \\ D \equiv a \,(\mathrm{mod}\, 8N)}} \left(\frac{D}{h} \right) |D|^{2s_0} L_D(1 + s_1, f) dt = \text{ main term } +$$

$$+ \mathbf{O}(Y^{1-\sigma_1+2\sigma_0} \left\{ \sqrt{h}(1 + |s_0| + |s_0 - s_1|)^2 (\log Yh)^\alpha \log(h \log Y) \right\})$$

for all $\sigma_0 \geq 0$, $0 \leq \sigma_1 < \frac{1}{2}$, $(ah, 2N) = 1$, $a \equiv 1 \,(\mathrm{mod}\, 4)$.

The "main term" here grows like $Y^{1+2\sigma_0}$. Thus, $A^\beta(\alpha)$ does not give an asymptotic formula unless $\sigma_1 > 0$.

$$C^\beta(\alpha): \quad \sum_{\substack{n \leq X \\ n_2 h \neq b^2}} \frac{a(n)}{n^{1+s_1}} \left(\frac{1}{Y} \int_1^Y f_t^\beta(nh, 2s_0; a) dt \right)$$

$$= \mathbf{O}(Y^{\frac{1}{2}+2\sigma_0} X^{\frac{1}{2}-\sigma_1} \frac{\sqrt{h}}{(2\sigma_1 - 1)}$$

$$\times (1 + |s_0| + |s_0 - s_1|)^2 (\log Yh)^\alpha \log(h \log Y))$$

for all $\sigma_0 \geq 0$, $0 < \sigma_1 < \frac{1}{2}$, $a \equiv 1 \,(\mathrm{mod}\, 4)$, and $(ah, 2N) = 1$, together with a similar estimate for the sum over $n \geq X$ and $\frac{1}{2} < \sigma_1 < 1$.

To produce the required estimate of $I(X, t)$, we need to know that for any $\lambda > 0$, $C^\beta(\lambda)$ holds.

After reviewing some basic estimates from [MM] in §5, we study the above statements in §6. We replace $C^\beta(\alpha)$ by a smoothed version, again denoted $C^\beta(\alpha)$. This version suffices for our application. Then, we show that $A^\beta(2)$ holds (Lemma 6.1), that $A^\beta(\alpha)$ implies $C^\beta(\alpha)$ (Proposition 6.3) and that $C^\beta(\alpha)$ implies $A^\beta(4\alpha/5)$ (Proposition 6.4).

This chapter is based on the preprint [M]. We note that Iwaniec [I] has found a beautiful method which proves the main result of [MM] with an improved error term of $\mathbf{O}(Y^\theta)$ with a $\theta < 1$. He accomplishes this by using Gauss sums to introduce an extra averaging. His method can be used to prove the asymptotic formula of Theorem 1.2 also and this is developed in [MS]. Recently, Friedberg and Hoffstein [FH] proved the non-vanishing result (but not the asymptotic formula) over any number field. Their method involves the Rankin-Selberg convolution and metaplectic Eisenstein series and is quite different from our techniques.

§2 The integrated Polya-Vinogradov estimate

We begin with the following well known integrated version of the Polya-Vinogradov estimate.

Lemma 2.1.   *Let $\beta = \pm 1$, and let χ be an even, nontrivial Dirichlet character modulo q. Then,*

$$\frac{1}{Y} \int_1^Y \sum_{0 < \beta D \le t} \chi(D) \, dt \ll \sqrt{q}.$$

Proof. This is due to Hua (see [BC, eqn.(7)]). $\blacksquare$

Lemma 2.2   *If $n_2 h \ne b^2$, and $(ah, 2N) = 1$, then*

$$\frac{1}{Y} \int_1^Y \sum_{\substack{|D| \le t \\ D \equiv a (\mathrm{mod}\, 8N)}} \left(\frac{D}{nh} \right) |D|^{2s_0} \, dt \ll (|s_0| + 1)(nh)^{\frac{1}{2}} Y^{2\sigma_0}.$$

Proof. We have

$$\sum_{\substack{|D| \le t \\ D \equiv a (\mathrm{mod}\, 8N)}} \left(\frac{D}{nh} \right) |D|^{2s_0}$$

$$= \frac{1}{\phi(8N)} \sum_{\psi (\mathrm{mod}\, 8N)} \bar{\psi}(a) \sum_{|D| \le t} \psi(D) \left(\frac{D}{nh} \right) |D|^{2s_0}$$

$$= \frac{1}{\phi(8N)} \sum_{\psi} \bar{\psi}(a) \sum_{0 < D \le t} \psi(D) \left(\frac{D}{nh} \right) |D|^{2s_0} \left(1 + \psi(-1) \left(\frac{-1}{nh} \right) \right).$$

Each character $\psi \left(\frac{\cdot}{nh} \right)$ is nontrivial as $\left(\frac{\cdot}{n_2 h} \right)$ is. Thus, we may apply Lemma 2.1 and partial summation to the inner sum to complete the proof. $\blacksquare$

Lemma 2.3   *For $\beta = \pm 1$, and $\sigma_0 \ge 0$, we have*

$$\sum_{\substack{0 < \beta D \le t \\ D \equiv a (\mathrm{mod}\, 8N), (D, n_2 h) = 1}} |D|^{2s_0} = \frac{1}{\phi(8N)} \frac{\phi(8N n_2 h)}{8N n_2 h} \frac{t^{1 + 2s_0}}{1 + 2s_0} + \mathbf{O}(t^{2\sigma_0} \mathbf{d}(n_2 h)(|s_0| + 1)).$$

This is proved by partial summation and some elementary estimates. The details are left to the reader.

As an immediate application, we state the following asymptotic estimates for average values of L-functions, the proof of which is left as an exercise. (We shall not need these in the remainder of the chapter.)

Proposition 2.4   *For $(ah, 2N) = 1$ and $\sigma_0 \geq 0$ we have*

$$\frac{1}{Y} \int_1^Y \sum_{\substack{|D| \leq t \\ D \equiv a \,(\mathrm{mod}\, 8N)}} \left(\frac{D}{h}\right) |D|^{2s_0} L_D(1 + s_1, f) dt$$

$$= C(s_0, s_1, 1, h) Y^{1+2s_0} + E(s_0, s_1, h, Y)$$

where if $\sigma_1 > 1$, $E(s_0, s_1, h, Y)$ is

$$\ll (|s_0| + 1) Y^{2\sigma_0} (h^{\frac{1}{2}} \zeta(\sigma_1) + \mathbf{d}(h)^2)$$

and if $\frac{1}{2} < \sigma_1 < 1$, it is

$$\ll Y^{1-\sigma_1+2\sigma_0} (\log Y)^{3(1-\sigma_1)} (|s_0| + 1) h^{\frac{1}{2}} ((1-\sigma_1)^{-1} + \left(\frac{Y}{\log Y}\right)^{\frac{1}{2}} |\Gamma(\sigma_1 - \frac{1}{2})|)).$$

We remark that using the results of §6, it is possible to refine the error terms.

To establish asymptotics for values of s_1 inside the critical strip, we shall need a more refined approach. Moreover, we shall need a substitute for Lemma 2.1 that holds for odd characters. These two problems are addressed in the next two sections.

§3 The main terms

We are interested in estimating the quantity

$$\frac{1}{Y} \int_1^Y \sum_{\substack{0 < \beta D \leq t \\ D \equiv a \,(\mathrm{mod}\, 8N)}} L_D(1 + s_1, f) dt$$

for $0 \leq \sigma_1 < 1/2$. However, it is important to treat a more general sum.

For $\sigma_0 \geq 0, 0 \leq \sigma_1 < \frac{1}{2}, a \equiv 1 (\mathrm{mod}\, 4), (ahj, 2N) = 1$, $(h, j) = 1$, $\beta = \pm 1$, we consider

$$\frac{1}{Y} \int_1^Y \sum_{\substack{0 < \beta D \leq t \\ D \equiv a \,(\mathrm{mod}\, 8N)}} \left(\frac{D}{h}\right) |D|^{2s_0} L_{Dj^2}(1 + s_1, f) \, dt.$$

Towards this end, for $D \equiv a (\mathrm{mod}\, 8N)$, consider the integral

$$\frac{1}{2\pi i} \int_{(2)} L_{Dj^2}(1 + s_1 + w, f) X^w \Gamma(w) dw.$$

This is

$$\sum_{(n,j)=1} \frac{a(n)}{n^{1+s_1}} \left(\frac{D}{n}\right) \exp(-n/X).$$

Moving the line of integration to $\operatorname{Re} s = -\eta$, with $0 < \eta < 1$, we obtain

$$\sum_{(n,j)=1} \frac{a(n)}{n^{1+s_1}} \left(\frac{D}{n}\right) \exp(-n/X)$$

$$= L_{Dj^2}(1+s_1,f) + \frac{1}{2\pi i} \int_{(-\eta)} L_{Dj^2}(1+s_1+w,f) X^w \Gamma(w) dw.$$

Writing $Dj^2 = D_0 \delta^2$ with D_0 a fundamental discriminant, we see that the integral is equal to

$$\frac{1}{2\pi i} \sum_{\substack{d|\delta^2 \\ j|\delta}} \frac{\tilde{\mu}(d,f)}{d^{1+s_1}} \int_{(-\eta)} L_{D_0}(1+s_1+w,f) \left(\frac{D_0}{d}\right) \left(\frac{X}{d}\right)^w \Gamma(w) dw.$$

Suppose now that $\eta > \frac{1}{2} + \sigma_1$. Then, on applying the functional equation, the integral becomes

$$\omega \chi_{D_0}(-N)\epsilon(D_0) \frac{1}{2\pi i} \sum_{\substack{d|\delta^2 \\ j|\delta}} \frac{\tilde{\mu}(d,f)}{d^{1+s_1}} \int_{(-\eta)} \sum \frac{\bar{a}(n)}{n^{1-s_1-w}} \left(\frac{D_0}{nd}\right) \times$$

$$\times (A|D_0|)^{-2(s_1+w)} \frac{\Gamma(1-s_1-w)}{\Gamma(1+s_1+w)} (X/d)^w \Gamma(w) dw.$$

Multiplying through by $|D|^{2s_0} \left(\frac{D}{h}\right)$, summing over

$$0 < \beta D \le t, \ D \equiv a(\operatorname{mod} 8N),$$

and integrating over t, we see that

$$\frac{1}{Y} \int_1^Y \sum_{\substack{0 < \beta D \le t \\ D \equiv a(\operatorname{mod} 8N)}} \left(\frac{D}{h}\right) |D|^{2s_0} L_{Dj^2}(1+s_1,f) \, dt$$

is equal to the sum of

$$\sum_{(n,j)=1} \frac{a(n)}{n^{1+s_1}} \frac{1}{Y} \int_1^Y \left(\sum_{\substack{0 < \beta D \le t \\ D \equiv a(\operatorname{mod} 8N)}} \left(\frac{D}{nh}\right) |D|^{2s_0} \, dt \right) \exp(-n/X)$$

and

$$- \beta \omega \epsilon(a) \left(\frac{a}{N}\right) A^{-2s_1} \sum_{\substack{\delta^2 \le Y \\ j|\delta \\ (\delta,2N)=1}} \bar{\epsilon}(\delta^2)\delta^{4s_0} \left(\frac{\delta^2}{h}\right) \sum_{d|\delta^2} \frac{\tilde{\mu}(d,f)}{d^{1+s_1}}$$

$$\times \frac{1}{2\pi i} \int_{(-\eta)} \sum \frac{\bar{a}(n)}{n^{1-s_1-w}} \left(\frac{1}{Y/\delta^2} \int_1^{Y/\delta^2} f_t^\beta(ndh, 2s_0 - 2s_1 - 2w; a\bar{\delta}^2) dt \right)$$

$$\times \frac{\Gamma(1-s_1-w)}{\Gamma(1+s_1+w)} (X/dA^2)^w \Gamma(w) dw.$$

Here, as usual, we are writing

$$f_t^\beta(ndh, 2s_0 - 2s_1 - 2w, a\bar{\delta}^2) \;=\; \sum_{\substack{0 < \beta D_0 \le t \\ D_0 \delta^2 \equiv a \,(\mathrm{mod}\, 8N)}} \left(\frac{D_0}{ndh}\right) |D_0|^{2s_0 - 2(s_1 + w)}.$$

We have now to determine the size of the sum and the integral. For the sum, we expect that the main term will come from the sum over those values of n for which $n_2 h = b^2$. This "main term" therefore is

$$\tilde{M}_\beta(s_0, s_1, j, h) \;=\; \sum_{\substack{n_2 h = b^2 \\ (n_2, j) = 1}} \frac{a(n)}{n^{1+s_1}} \left(\frac{a}{n_1}\right) \left(\frac{1}{Y} \int_1^Y \sum_{\substack{0 < \beta D \le t \\ D \equiv a \,(\mathrm{mod}\, 8N) \\ (D, n_2 h) = 1}} |D|^{2s_0} \, dt\right) \exp(-n/X).$$

When there is no possibility of confusion, we shall simply write

$$M \;=\; \tilde{M}_\beta(s_0, s_1, j, h).$$

It is more convenient to work with the series

$$M_\beta(s_0, s_1, j, h) \;\overset{\mathrm{def}}{=}\; \sum_{\substack{n_2 h = b^2 \\ (n_2, j) = 1}} \frac{a(n)}{n^{1+s_1}} \left(\frac{a}{n_1}\right) \left(\frac{1}{Y} \int_1^Y \sum_{\substack{0 < \beta D \le t \\ D \equiv a \,(\mathrm{mod}\, 8N) \\ (D, n_2 h) = 1}} |D|^{2s_0} \, dt\right).$$

For $\sigma \ge 0$, and $(d, 2ND) = 1$, define

$$F_d(s, f, D) \;=\; \sum_{\substack{m_2 d = b^2 \\ (m, D) = 1}} \frac{a(m)}{m^{1+s}} \left(\frac{a}{m_1}\right).$$

In terms of this function, we see that

$$M_\beta(s_0, s_1, j, h) \;=\; \frac{1}{Y} \int_1^Y \sum_{\substack{0 < \beta D \le t \\ D \equiv a \,(\mathrm{mod}\, 8N) \\ (D, h) = 1}} |D|^{2s_0} F_h(s_1, f, jD)\, dt.$$

The two series $\tilde{M}_\beta$ and M_β are related by the following estimate.

Lemma 3.1   *We have*

$$\tilde{M}_\beta(s_0, s_1, j, h) \;=\; M_\beta(s_0, s_1, j, h) + \mathbf{O}\Big(\prod_{p \mid j} \Big(1 + \frac{1}{p^{3/4}}\Big)^3 X^{-\sigma_1 - \frac{1}{8}} Y^{1 + 2\sigma_0}\Big).$$

uniformly in h.

The proof will be given later in this section.

As for the integral, we again expect that the "main term" will come from the sum of those values of n where $n_2 dh$ is a perfect square. The "main term" is therefore

$$-\beta\omega\epsilon(a)\left(\frac{a}{N}\right)A^{-2s_1}\sum_{\substack{\delta^2\leq Y\\ j\mid\delta\\ (\delta,2N)=1}}\bar{\epsilon}(\delta^2)\delta^{4s_0}\left(\frac{\delta^2}{h}\right)$$

$$\times\sum_{d\mid\delta^2}\frac{\tilde{\mu}(d,f)}{d^{1+s_1}}\frac{1}{2\pi i}\int_{(-\eta)}\sum_{n_2 dh=b^2}\frac{\bar{a}(n)}{n^{1-s_1-w}}\left(\frac{a}{n_1}\right)$$

$$\times\left(\frac{1}{Y/\delta^2}\int_1^{Y/\delta^2}\sum_{\substack{0<\beta D_0\leq t\\ D_0\delta^2\equiv a(\bmod 8N)\\ (D_0,2Nn_2 dh)=1}}|D_0|^{-2(s_1+w-s_0)}\,dt\right.$$

$$\times\left.\frac{\Gamma(1-s_1-w)}{\Gamma(1+s_1+w)}(X/dA^2)^w\Gamma(w)dw.\right.$$

For $\eta\geq\sigma_1$, the "main term" may be written in terms of the function F_d as follows:

$$N = -\beta\omega\epsilon(a)\left(\frac{a}{N}\right)A^{-2s_1}\sum_{\substack{\delta^2\leq Y\\ j\mid\delta\\ (\delta,2N)=1}}\bar{\epsilon}(\delta^2)\delta^{4s_0}\left(\frac{\delta^2}{h}\right)\sum_{d\mid\delta^2}\frac{\tilde{\mu}(d,f)}{d^{1+s_1}}$$

$$\times\frac{1}{Y/\delta^2}\int_1^{Y/\delta^2}\sum|D_0|^{2s_0-2s_1}$$

$$\times\left(\frac{1}{2\pi i}\int_{(-\eta)}\left(\frac{X}{dA^2|D_0|^2}\right)^w F_{dh}(-s_1-w,\bar{f},D_0)\frac{\Gamma(1-s_1-w)}{\Gamma(1+s_1+w)}\Gamma(w)dw\right)dt.$$

If $0<\sigma_1<\frac{1}{8}$, we move the line of integration to the right, and get a residue at $w=0$ equal to

$$N_\beta(s_0,s_1,j,h)\overset{\text{def}}{=}\beta\omega\epsilon(a)\left(\frac{a}{N}\right)A^{-2s_1}\frac{\Gamma(1-s_1)}{\Gamma(1+s_1)}\sum_{\substack{\delta^2\leq Y\\ j\mid\delta\\ (\delta,2N)=1}}\bar{\epsilon}(\delta^2)\delta^{4s_0}\left(\frac{\delta^2}{h}\right)\sum_{d\mid\delta^2}\frac{\tilde{\mu}(d,f)}{d^{1+s_1}}$$

$$\times\left(\frac{1}{Y/\delta^2}\int_1^{Y/\delta^2}\sum_{\substack{0<\beta D_0\leq t\\ D_0\delta^2\equiv a(\bmod 8N)}}|D_0|^{2(s_0-s_1)}F_{dh}(-s_1,\bar{f},D_0)dt\right).$$

We also set

$$N_\beta(s_0,s_1,j,h)=0\qquad\text{if }\frac{1}{8}<\sigma_1<\frac{1}{2}.$$

Remark. The point here is that the residue is insignificant if σ_1 is bounded away from zero. The choice of $1/8$ as a breaking point was only for convenience, and any other point will work just as well.

Lemma 3.2 *For $0 \leq \sigma_1 < \frac{1}{8}$, we have*

$$N \; = \; N_\beta(s_0, s_1, j, h) \; + \; O(Y^{\frac{5}{8}+2\sigma_0} X^{\frac{3}{16}-\sigma_1}).$$

For $\frac{1}{8} \leq \sigma_1 < \frac{1}{2}$, we have

$$N \; \ll \; Y^{1+2(\sigma_0-\sigma_1+\eta)} X^{-\eta} |\Gamma(-\eta)|$$

for any $\sigma_1 \leq \eta < 1$.

The proof will be given later in the section.

We expect that

$$\frac{1}{Y} \int_1^Y \sum_{\substack{0<\beta D \leq t \\ D \equiv a \,(\mathrm{mod}\, 8N)}} \left(\frac{D}{h}\right) |D|^{2s_0} L_{Dj^2}(1+s_1, f) dt$$

is asymptotic to

$$M_\beta(s_0, s_1, j, h) \; + \; N_\beta(s_0, s_1, j, h).$$

It will be necessary to have estimates on the growth of these functions. This will be based on a study of the function $F_d(s, f, D)$ introduced above.

Analysis of $F_d(s, f, D)$

Let us define

$$B_p(s) \; = \; \sum_{\ell=0}^{\infty} \frac{a(p^{2\ell})}{p^{2\ell s}}.$$

Let us also define the function

$$F_{d_0}(s, f) = \sum \frac{a(d_0 n^2)}{(d_0 n^2)^s}$$

for $Re(s) > 1$. Then, by Exercise 3,

$$F_{d_0}(s, f) \; = \; L(2s, Sym^2)\zeta(4s-2)^{-1} \prod_{p|d_0} \left(1 - \frac{\epsilon(p)p}{p^{2s}}\right)\left(1 - \frac{1}{p^{4s-2}}\right)^{-1}\left(\frac{1+\epsilon(p)p}{p^s}\right).$$

and this gives the analytic continuation of $F_{d_0}(s)$ for $Re(s) > 3/4$.

Now, we have the relation

$$
F_d(s, f, D) = \left(\sum_{\substack{p|m_1 \Rightarrow p|2N \\ (m_1, D)=1}} \frac{a(m_1)}{m_1^{1+s}} \left(\frac{a}{m_1}\right) \right) \left(\sum_{\substack{m_2 d = b^2 \\ (m_2, D)=1}} \frac{a(m_2)}{m_2^{1+s}} \right).
$$

The first factor is entire for $\sigma > -\frac{1}{2}$, and uniformly bounded for $\sigma \geq -3/16$. Now, let us write $d = d_0 d_1^2$ with d_0 squarefree. As for the second factor, we see then that it is equal to

$$
\sum_{(n, 2ND)=1} \frac{a(n^2 d_0)}{(n^2 d_0)^{1+s}} = F_{d_0}(1+s, f) \prod_{p|2ND} \left(\sum_{\ell=0}^{\infty} \frac{a(p^{2\ell})}{p^{2\ell(1+s)}} \right)^{-1}.
$$

Suppose that $(D, 2N) = 1$. Then, we can write

$$
F_d(s, f, D) = F_{d_0}(1+s, f) G(s, f) \prod_{p|D} B_p(1+s)^{-1}
$$

where we have set

$$
G(s, f) = \left(\sum_{\substack{p|m \Rightarrow p|2N \\ (m, D)=1}} \frac{a(m)}{m^{1+s}} \left(\frac{a}{m}\right) \right) \prod_{p|2N} B_p(1+s)^{-1}.
$$

It is clear that $G(s, f)$ is analytic for $Re(s) > -\frac{1}{2}$. Combining this with what was stated earlier, it follows that $F_d(s, f, D)$ is analytic for $Re(s) > -1/4$.

Lemma 3.3 *Write $d = d_0 d_1^2$ with d_0 squarefree, and suppose that $(d, 2ND) = 1$ and $(D, 2N) = 1$. Then, for $\sigma \geq -3/16$, and an absolute constant $c_1 > 0$,*

$$
F_d(s, f, D) \ll c_1^{\nu(d_0)} (d_0)^{-\sigma - \frac{1}{2}} |L(2s+2, Sym^2) \zeta(2+4s)^{-1}| \prod_{p|D} (1 + \frac{1}{p^{1+2\sigma}})^3.
$$

Proof. The proof of [MM, Lemma 13] shows that

$$
F_d(1+s, f) \ll c_2^{\nu(d_0)} (d_0)^{-\sigma - \frac{1}{2}} |L(2s+2, Sym^2) \zeta(2+4s)^{-1}|
$$

for an absolute constant $c_2 > 0$ and $\sigma > -1/4$. It is easily proved that for $\sigma \geq -3/16$, (say),

$$
\left(\sum_{\ell=0}^{\infty} \frac{a(p^{2\ell})}{p^{2\ell(1+s)}} \right)^{-1} \ll \left(1 + \frac{1}{p^{1+2\sigma}} \right)^3 \left(1 - \frac{1}{p^{2+4\sigma}} \right)^{-1}
$$

and this is

$$
\ll \left(1 + \frac{1}{p^{1+2\sigma}} \right)^3.
$$

It follows that the desired result holds with a possibly larger value of c_2.

Remark. It should be clear that it is only essential that σ be larger than and bounded away from $-1/4$.

Now we give the proofs of results stated earlier.

Proof of Lemma 3.1. We have

$$M = \frac{1}{2\pi i}\int_{(2)}\sum_{\substack{n_2 h=b^2\\(n_2,j)=1}}\frac{a(n)}{n^{1+s_1+w}}\left(\frac{a}{n_1}\right)\left(\frac{1}{Y}\int_1^Y\sum_{\substack{0<\beta D\le t\\D\equiv a(\bmod 8N)\\(D,n_2 h)=1}}|D|^{2s_0}\,dt\right)X^w\Gamma(w)dw.$$

Moving the line of integration to $Re(w)=-\sigma_1-\frac18$, we see that

$$M = M_\beta(s_0,s_1,j,h) + \frac{1}{2\pi i}\int_{(-\sigma_1-\frac18)}M_\beta(s_0,s_1+w,j,h)X^w\Gamma(w)dw.$$

By Lemma 3.3, we see that the integral is

$$\ll \prod_{p|j}\left(1-\frac{1}{p^{3/4}}\right)^{-3}X^{-\sigma_1-\frac18}Y^{1+2\sigma_0}|\Gamma(-\sigma_1-\frac18)|.$$

This proves the result.

Proof of Lemma 3.2. Suppose first that $0\le\sigma_1<\frac18$. By definition,

$$N - N_\beta(s_0,s_1,j,h)$$

is equal to

$$-\beta w\epsilon(a)\left(\frac{a}{N}\right)A^{-2s_1}\sum_{\substack{\delta^2\le Y\\j|\delta}}\bar\epsilon(\delta^2)\delta^{4s_0}\left(\frac{\delta^2}{h}\right)\sum_{d|\delta^2}\frac{\tilde\mu(d,f)}{d^{1+s_1}}$$

$$\frac{1}{Y/\delta^2}\int_1^{Y/\delta^2}\sum|D_0|^{2s_0-2s_1}$$

$$\left(\frac{1}{2\pi i}\int_{(\eta)}\left(\frac{X}{dA^2|D_0|^2}\right)^w F_{dh}(-s_1-w,\bar f,D_0)\frac{\Gamma(1-s_1-w)}{\Gamma(1+s_1+w)}\Gamma(w)dw\right)dt$$

for some $\frac14>\eta>0$. Using the factorization

$$F_d(s,f,D) = F_{d_0}(1+s,f)G(s,f)\prod_{p|D}B_p(1+s)^{-1}$$

described earlier, we see that

$$\frac{1}{Y/\delta^2}\int_1^{Y/\delta^2}\sum|D_0|^{2s_0-2s_1}$$
$$\left(\frac{1}{2\pi i}\int_{(\eta)}\left(\frac{X}{dA^2|D_0|^2}\right)^w F_{dh}(-s_1-w,\bar{f},D_0)\frac{\Gamma(1-s_1-w)}{\Gamma(1+s_1+w)}\Gamma(w)dw\right)dt$$

is equal to

$$\frac{1}{2\pi i}\int_{(\eta)}F_{d_0}(1-s_1-w,\bar{f})G(-s_1-w,\bar{f})$$
$$\left(\frac{1}{Y/\delta^2}\int_1^{Y/\delta^2}\sum_{0<\beta D_0\le t}|D_0|^{2s_0-2s_1-2w}\prod_{p|D_0}B_p(1-s_1-w)^{-1}dt\right)$$
$$\left(\frac{X}{dA^2}\right)^w\frac{\Gamma(1-s_1-w)}{\Gamma(1+s_1+w)}\Gamma(w)dw.$$

By an identity stated earlier, we get an estimate

$$|B_p(1+s)|^{-1}\ll\left(1+\frac{1}{p^{2\sigma+1}}\right)^3\left(1-\frac{1}{p^{4\sigma+2}}\right)^{-1}.$$

Thus,

$$\sum_{0<\beta D_0\le t}|D_0|^{2s_0-2s_1-2w}\prod_{p|D_0}B_p(1-s_1-w)^{-1}$$
$$\ll\sum_{0<\beta D_0\le t}|D_0|^{2\sigma_0-2\sigma_1-2\eta}\prod_{p|D_0}\left(1+\frac{1}{p^{1-2\sigma_1-2\eta}}\right)^3\left(1-\frac{1}{p^{2-4\sigma_1-4\eta}}\right)^{-1}$$

and choosing $\eta=\frac{3}{16}-\sigma_1$ (say), this is

$$\ll\sum_{0<\beta D_0\le t}|D_0|^{2\sigma_0-\frac{3}{8}}\sigma_{-5/8}(D_0)^3\ll t^{2\sigma_0+\frac{5}{8}}.$$

Also,

$$F_{d_0}(1-s_1-w,\bar{f})\ll|L(2-2s_1-2w,Sym^2)\zeta(2-4s_1-4w)^{-1}|$$
$$\times\left\{\prod_{p|d_0}\left(1+\frac{p}{p^{2-2\sigma_1-2\eta}}\right)\left(1-\frac{1}{p^{2-4\sigma_1-4\eta}}\right)^{-1}p^{\sigma_1+\eta}\right\}.$$

Inserting all this into the integral, we see that the entire expression that we are trying to estimate is

$$\ll \sum_{\substack{\delta^2 \leq Y \\ j|\delta}} \delta^{4\sigma_0} \sum_{d|\delta^2} \frac{d^{\frac{1}{2}}\mathbf{d}(d)}{d^{1+\sigma_1}}$$

$$\times \prod_{p|d_0} \left\{ \left(1 + \frac{p}{p^{13/8}}\right)\left(1 - \frac{1}{p^{5/4}}\right)^{-1} p^{3/16} \right\} \left(\frac{Y}{\delta^2}\right)^{2\sigma_0+\frac{5}{8}} \left(\frac{X}{d}\right)^{\frac{3}{16}-\sigma_1}$$

which is

$$\ll Y^{2\sigma_0+\frac{5}{8}} X^{\frac{3}{16}-\sigma_1} \sum_{\substack{\delta^2 \leq Y \\ j|\delta}} \delta^{-5/4} \sum_{d|\delta^2} \frac{d^{\frac{1}{2}}\mathbf{d}(d)}{d^{19/16}} d^{3/16} \sigma_{-5/8}(d)$$

and simplifying, we see that this is

$$\ll Y^{2\sigma_0+\frac{5}{8}} X^{\frac{3}{16}-\sigma_1}.$$

For $\frac{1}{2} > \sigma_1 > \frac{1}{8}$, Lemma 3.3 implies that for $\sigma_1 \leq \eta < 1$,

$$N \ll Y^{1+2(\sigma_0-\sigma_1+\eta)} X^{-\eta} |\Gamma(-\eta)|.$$

This proves the lemma.

Lemma 3.4 Let $(h, 2N) = 1$. We have for $\sigma_1 \neq 0$, and $-\sigma_1 + \frac{1}{8} > c \geq -\sigma_1$

$$\frac{1}{2\pi i} \int_{(c)} N_\beta(s_0, s_1 + w, j, h) X^w \Gamma(w) dw \ll |\Gamma(c)| Y^{1+2(\sigma_0-\sigma_1-c)} X^c.$$

Proof. By definition, the integral of the lemma is equal to

$$\beta w \epsilon(a) \left(\frac{a}{N}\right) \sum_{\substack{\delta^2 \leq Y \\ j|\delta}} \bar{\epsilon}(\delta^2) \delta^{4s_0} \left(\frac{\delta^2}{h}\right) \sum_{d|\delta^2} \frac{\tilde{\mu}(d, f)}{d^{1+s_1}} A^{-2s_1}$$

$$\frac{1}{Y/\delta^2} \int_1^{Y/\delta^2} \sum |D_0|^{2s_0-2s_1}$$

$$\left(\frac{1}{2\pi i} \int_{(c)} \left(\frac{X}{dA^2|D_0|^2}\right)^w F_{dh}(-s_1 - w, \bar{f}, D_0) \frac{\Gamma(1 - s_1 - w)}{\Gamma(1 + s_1 + w)} \Gamma(w) dw\right) dt.$$

By Lemma 3.3, we see that uniformly in d and h, the inner integral is

$$\ll |\Gamma(c)| \left(\frac{X}{dA^2|D_0|^2}\right)^c \sigma_{-7/8}(D_0)^3.$$

Hence, the sum over D_0 is

$$\ll \sum |D_0|^{2(\sigma_0 - \sigma_1)} \left(\frac{X}{dA^2|D_0|^2} \right)^c \sigma_{-7/8}(D_0)^3 |\Gamma(c)|$$

which is

$$\ll \left(\frac{X}{dA^2} \right)^c t^{1+2(\sigma_0 - \sigma_1 - c)} |\Gamma(c)|.$$

Inserting this into the big expression, we see that

$$\frac{1}{2\pi i} \int_{(-\sigma_1)} N_\beta(s_0, s_1 + w, j, h) X^w \Gamma(w) dw$$

$$\ll \sum_{\substack{\delta^2 \le Y \\ j|\delta}} \delta^{4\sigma_0} \sum_{d|\delta^2} \frac{|\tilde{\mu}(d,f)|}{d^{1+\sigma_1}} \left(\frac{X}{d} \right)^c \left(\frac{Y}{\delta^2} \right)^{1+2(\sigma_0 - \sigma_1 - c)} |\Gamma(c)|$$

$$\ll |\Gamma(c)| Y^{1+2(\sigma_0 - \sigma_1 - c)} X^c \sum_{\delta^2 \le Y} \delta^{-2+4(\sigma_1 + c)} \sum_{d|\delta^2} \frac{d(d)}{d^{\frac{1}{2} + \sigma_1 + c}}$$

$$\ll |\Gamma(c)| Y^{2(\sigma_0 - \sigma_1 - c) + 1} X^c.$$

We conclude this section with an estimate that we shall use in §7.

Lemma 3.5 *We have*

$$\tilde{M}_\beta(0,0,1,1) = \frac{1}{2} C(0,0,1,1,) Y + \mathbf{O}(YX^{-1/8}) + \mathbf{O}((\log X)^3).$$

Proof. Inserting the asymptotic formula provided by Lemma 2.3 into the integral, we see that it is equal to

$$\sum_{n_2 = b^2} \frac{a(n)}{n} \left(\frac{a}{n_1} \right) \left\{ \frac{1}{Y} \int_1^Y \left(\frac{1}{8N} \frac{\phi(n_2)}{n_2} t + \mathbf{O}(d(n_2)) \right) dt \right\} \exp(-n/X).$$

The error term contributes an amount which is

$$\ll \sum_{n_1} \frac{|a(n_1)|}{n_1} \sum_b \frac{|a(b^2)|}{b^2} d(b^2) \exp(-b^2/X)$$

$$\ll \prod_{p|2N} \left\{ 1 + \frac{|a(p)|}{p} + \frac{|a(p^2)|}{p^2} + \cdots \right\} \sum_b \frac{d(b^2)}{b} \exp(-b^2/X).$$

The first factor depends only on N and can be absorbed into the implied constant. For the sum over b, we see by standard estimates that it is

$$\ll (\log X)^3.$$

Now consider the contribution of the main term. We see that it is equal to

$$\frac{Y}{16N} \sum_{n_2 = b^2} \frac{a(n)}{n} \left(\frac{a}{n_1}\right) \frac{\phi(n_2)}{n_2} \exp(-n/X)$$

which is

$$\frac{Y}{16N} \frac{1}{2\pi i} \int_{(2)} \tilde{F}_1(w, f, 1) X^w \Gamma(w) dw$$

Here, for $(d, 2ND) = 1$, we have set

$$\tilde{F}_d(s, f, D) \;=\; \sum_{\substack{m_2 d = b^2 \\ (m, D) = 1}} \frac{a(m)}{m^{1+s}} \left(\frac{a}{m_1}\right) \frac{\phi(m_2 d)}{m_2 d}.$$

Using the properties of $\tilde{F}_d$ given in Exercise 4, and moving the line of integration to the left, we see that this is

$$\frac{Y}{16N} \left\{ \sum_{n_2 = b^2} \frac{a(n)}{n} \left(\frac{a}{n_1}\right) \frac{\phi(n_2)}{n_2} + \frac{1}{2\pi i} \int_{(-1/8)} \tilde{F}_1(w, f, 1) X^w \Gamma(w) dw \right\}.$$

Now applying the estimate for $\tilde{F}_1$ of Exercise 4, we see that this is

$$\frac{Y}{16N} \tilde{F}_1(0, f, 1) \;+\; \mathbf{O}(YX^{-1/8})$$

as required.

§4 Estimates for real character sums

In this section, we develop a substitute for Lemma 2.1 in the case of odd Dirichlet characters.

Lemma 4.1 *Let χ be a fixed Dirichlet character and h an integer. Then,*

$$\sum_{n \leq X} {}^* n |L(1, \chi\left(\frac{\cdot}{hn}\right))|^2 \;\ll\; X^2$$

where the sum over n only includes values such that $\chi\left(\frac{\cdot}{hn}\right)$ is a nontrivial character, and the implied constant depends only on the conductor of χ.

Proof. This follows by a small modification in the proof of Jutila [J, Theorem 3] and partial summation. Indeed, we have

$$L\left(1, \chi\left(\frac{\cdot}{hn}\right)\right) = \sum_{m \leq X} \frac{\chi(m)}{m}\left(\frac{m}{hn}\right) + \mathbf{O}\left(\frac{\log X}{\sqrt{X}}\right)$$

for fixed χ and all $n \leq X$ such that $\chi\left(\frac{\cdot}{hn}\right) \neq 1$. It follows that

$$\sum_{n \leq X}{}^{*} n \left|L\left(1, \chi\left(\frac{\cdot}{hn}\right)\right)\right|^2 = \sum_{n \leq X}{}^{*} n \left|\sum_{m \leq X} \frac{\chi(m)}{m}\left(\frac{m}{hn}\right)\right|^2 + \mathbf{O}(X^{3/2}(\log X)^2).$$

Expanding the first term, we see that it is the sum of

$$A = \sum_{n \leq X}{}^{*} n \sum_{\substack{r \leq X \\ (r, hn)=1}} \frac{\overline{\chi(r^2)}}{r^2} \sum_{m | r^2} \chi(m)^2$$

and

$$B = \sum_{\substack{m_1, m_2 \leq X \\ m_1 m_2 \neq b^2}} \frac{\chi(m_1 \overline{m_2})}{m_1 m_2} \sum_{n \leq X}{}^{*} n \left(\frac{m_1 m_2}{hn}\right).$$

It is clear that $A = \mathbf{O}(X^2)$. (In fact, it is asymptotic to cX^2 for some $c \neq 0$). As for B, suppose χ is a character mod r. We note that

$$\left| \sum_{\substack{n \leq X \\ \chi(d) = \left(\frac{d}{hn}\right)}} n\left(\frac{d}{n}\right) \right| \leq X^{3/2} \sum_{\substack{n \leq X \\ p | n \Rightarrow p | r}} \frac{1}{\sqrt{n}} \ll X^{3/2} \prod_{p | r}\left(1 - p^{-1/2}\right)^{-1}.$$

Let us set

$$S(d) = \sum_{n \leq X} n\left(\frac{d}{n}\right).$$

As r is fixed, we deduce that

$$\sum_{n \leq X}{}^{*} n\left(\frac{d}{hn}\right) = \left(\frac{d}{h}\right) S(d) + \mathbf{O}(X^{3/2}).$$

Therefore,

$$B \ll X^{3/2}(\log X)^2 + \sum_{\substack{m \leq X^2 \\ m \neq b^2}} \frac{\mathbf{d}(m)}{m}|S(m)|.$$

If we set

$$T(Z, d) = \sum_{n \leq Z}\left(\frac{d}{n}\right), \quad \text{then} \quad S(d) = XT(X, d) - \sum_{n \leq X-1} T(n, d)$$

and [J, Theorem 1] (see also [MM, Lemma 1]) asserts that

$$\sum_{\substack{d \leq W \\ d \neq b^2}} |T(Z, d)|^2 \ll ZW(\log W)^2.$$

By partial summation, it follows that

$$\sum_{\substack{d \leq W \\ d \neq b^2}} \frac{1}{d} |T(Z,d)|^2 \ll Z(\log W)^3.$$

Thus, we see that for $1 \leq n \leq X$, we have

$$\sum_{\substack{m \leq X^2 \\ m \neq b^2}} \frac{d(m)}{m} |T(n,m)| \leq \left(\sum_{m \leq X^2} \frac{d(m)^2}{m} \right)^{1/2} \left(\sum_{\substack{m \leq X^2 \\ m \neq b^2}} \frac{1}{m} |T(n,m)|^2 \right)^{1/2}$$

$$\ll (\log X)^2 n^{1/2} (\log X)^{3/2} = n^{1/2} (\log X)^{7/2}.$$

It follows that

$$\sum_{\substack{m \leq X^2 \\ m \neq b^2}} \frac{d(m)}{m} |S(m)| \ll X^{3/2} (\log X)^{7/2}.$$

It follows that $B = \mathbf{O}(X^{3/2}(\log X)^{7/2})$ and this proves the lemma.

Recall that we have set

$$\log^+ a = \begin{cases} \log a & \text{if } a > 1 \\ 1 & \text{otherwise} \end{cases}.$$

Lemma 4.2. *Let χ be a fixed Dirichlet character and $(h, 2N) = 1$. Then*

$$\sum_{n \leq X}^{*} \left| \frac{1}{Y} \int_0^Y \sum_{0 < D \leq t} \chi(D) \left(\frac{D}{hn} \right) dt \right|^2 \ll hX^2 (\log^+ \frac{Xh}{Y})^2$$

where the sum over n is as above.

Proof. Suppose first that $\psi = \chi \left(\frac{\cdot}{hn} \right)$ is a primitive character, $\bmod\, q$ (say). Polya has derived the expression

$$\sum_{0 < D \leq t} \chi(D) \left(\frac{D}{hn} \right) = g(\psi) \sum_{0 < |m| \leq H} \bar{\psi}(m) \left(\frac{1 - e^{-2\pi i mt/q}}{2\pi i m} \right) + \mathbf{O}(1 + \frac{q \log q}{H}).$$

(See [BC] or [MV].) Here $g(\psi)$ is the Gauss sum attached to ψ. Integrating with respect to t, we deduce that

$$\frac{1}{Y} \int_0^Y \left(\sum_{0 < D \leq t} \chi(D) \left(\frac{D}{hn} \right) \right) dt$$

$$= g(\psi) \sum_{0 < |m| \leq H} \frac{\bar{\psi}(m)}{2\pi i m} + \mathbf{O}(1 + \frac{q \log q}{H}) + \frac{q g(\psi)}{4\pi^2 Y} \sum_{0 < |m| \leq H} \frac{\bar{\psi}(m)}{m^2} \left(1 - e^{-2\pi i m Y/q} \right).$$

Now,

$$\sum_{0<|m|\leq H} \frac{\bar{\psi}(m)}{2\pi i m} = \frac{1}{2\pi i}(1 - \bar{\psi}(-1))(L(1,\bar{\psi}) + \mathbf{O}(\frac{q}{H})).$$

Thus, letting $H \longrightarrow \infty$, we find that

$$\frac{1}{Y}\int_0^Y \left(\sum_{0<D\leq t} \chi(D)\left(\frac{D}{hn}\right)\right) dt$$

is equal to

$$g(\psi)\frac{1}{2\pi i}(1 - \bar{\psi}(-1))L(1,\bar{\psi}) + \mathbf{O}(1) + \frac{qg(\psi)}{4\pi^2 Y}(1 + \bar{\psi}(-1))L(2,\bar{\psi})$$

$$- \frac{qg(\psi)}{4\pi^2 Y}\sum_{m=1}^\infty \frac{\bar{\psi}(m)}{m^2}\left(e^{-2\pi imY/q} + \bar{\psi}(-1)e^{2\pi imY/q}\right).$$

If $\psi(-1) = -1$, then

$$e^{-2\pi imY/q} + \bar{\psi}(-1)e^{2\pi imY/q} = -2i\sin\frac{2\pi mY}{q} \ll \frac{mY}{q}.$$

Hence, in this case,

$$\frac{1}{Y}\int_0^Y \left(\sum_{0<D\leq t} \chi(D)\left(\frac{D}{hn}\right)\right) dt$$

$$= \frac{g(\psi)}{\pi i}L(1,\bar{\psi}) + \mathbf{O}(1) + \mathbf{O}(\frac{q^{3/2}}{Y}\sum_{m=1}^\infty \frac{1}{m^2}\min(1,\frac{mY}{q}))$$

and this is equal to

$$\frac{g(\psi)}{\pi i}L(1,\bar{\psi}) + \mathbf{O}(1) + \mathbf{O}(q^{\frac{1}{2}}\log^+ \frac{q}{Y}).$$

If $\psi(-1) = +1$, then by Lemma 2.1

$$\frac{1}{Y}\int_1^Y \left(\sum_{0<D\leq t} \chi(D)\left(\frac{D}{hn}\right)\right) dt \ll \sqrt{q}.$$

Hence, in all cases,

$$\left|\frac{1}{Y}\int_0^Y \left(\sum_{0<D\leq t} \chi(D)\left(\frac{D}{hn}\right)\right) dt\right|^2 \ll q|L(1,\bar{\psi})|^2 + \mathbf{O}(1) + \mathbf{O}(q(\log^+ \frac{q}{Y})^2).$$

Note that q is $\mathbf{O}(hn)$. Hence, summing over those n such that $\chi\left(\frac{\cdot}{hn}\right) \neq 1$, we get

$$\sum_{n \leq \min(X,Y/h)}^{*} \left| \frac{1}{Y} \int_0^Y \left(\sum_{0 < D \leq t} \chi(D) \left(\frac{D}{hn} \right) \right) dt \right|^2$$

$$\ll \sum_{n \leq X}^{*} hn |L(1, \bar{\chi}\left(\frac{\cdot}{hn}\right))|^2 + \mathbf{O}(hX^2)$$

$$\ll hX^2.$$

In the last step, we used the estimate of Lemma 4.1. If $Xh \geq Y$, then

$$\sum_{Y/h < n \leq X} \ll \sum_{n \leq X}^{*} hn |L(1, \chi\left(\frac{\cdot}{hn}\right))|^2 + \sum_{Y < n \leq X} hn (\log \frac{hn}{Y})^2$$

and this is

$$\ll hX^2 + hX^2 (\log \frac{Xh}{Y})^2.$$

If ψ is not primitive, we let ψ^* denote the primitive character mod q^* say, that induces ψ. Let us write $q = q^*r$. Then, we have

$$\frac{1}{Y} \int_0^Y \left(\sum_{0 < D \leq t} \chi(D) \left(\frac{D}{hn} \right) \right) dt = \frac{1}{Y} \int_0^Y \left(\sum_{\substack{0 < D \leq t \\ (D,r)=1}} \psi^*(D) \right) dt$$

and the latter is equal to

$$\sum_{d|r} \mu(d) \psi^*(d) \frac{1}{Y/d} \int_0^{Y/d} \left(\sum_{0 < D \leq t} \psi^*(D) \right) dt.$$

Thus,

$$\sum_{n \leq X}^{*} \left| \frac{1}{Y} \int_0^Y \left(\sum_{0 < D \leq t} \chi(D) \left(\frac{D}{hn} \right) \right) dt \right|^2$$

$$\ll \sum_{n \leq X}^{*} \mathbf{d}(r)^2 \sum_{d|r} \left| \frac{1}{Y/d} \int_0^{Y/d} \left(\sum_{0 < D \leq t} \psi^*(D) \right) dt \right|^2.$$

The right side above is

$$\ll \sum_{n \leq X}^{*} \mathbf{d}(r)^2 \sum_{d|r} \left(q^* |L(1, \overline{\psi^*})|^2 + \mathbf{O}(1) + \mathbf{O}(q^* (\log^+ \frac{q^*}{Y})^2) \right).$$

Since
$$|L(1, \overline{\psi^*})| \ll \frac{r}{\phi(r)} L(1, \overline{\psi})$$

it follows that the above is

$$\ll \sum_{n \leq X} {}^* q|L(1, \overline{\psi})|^2 \frac{\mathbf{d}(r)^3 r}{\phi(r)^2} + \mathbf{O}\left(\sum_{n \leq X} \mathbf{d}(r)^3\right) + \mathbf{O}\left(\sum_{n \leq X} \mathbf{d}(r)^3 q^* (\log^+ \frac{q^*}{Y})^2\right).$$

Since $\mathbf{d}(r) \ll r^\epsilon$ and $\phi(r) \geq r/(\log \log r)$, the quantity $\mathbf{d}(r)^3 r/\phi(r)^2$ is bounded. Thus, by Lemma 4.1, the first term above is $\ll hX^2$. The second $\mathbf{O}$ term above is clearly $\ll X(\log X)^7$. Finally, as $\mathbf{d}(r)^3 q^* \leq q$, the final $\mathbf{O}$ term is $\ll hX^2(\log^+ Xh/Y)^2$.

Lemma 4.3. *With notation as above and $\sigma \geq 0$, we have*

$$\sum_{n \leq X} {}^* \left| \frac{1}{Y} \int_1^Y \left(\sum_{0 < D \leq t} \chi(D) \left(\frac{D}{hn}\right) \right) t^s dt \right|^2 \ll h(1 + |s|^2) Y^{2\sigma} X^2 (\log^+ \frac{Xh}{Y})^2$$

Proof. This follows from Lemma 4.2 by integration by parts and an application of the Cauchy-Schwarz inequality.

Lemma 4.4 *With notation as above and $\sigma \geq 0$, we have*

$$\sum_{n \leq X} {}^* \left| \frac{1}{Y} \int_1^Y \left(\sum_{0 < D \leq t} \chi(D) \left(\frac{D}{hn}\right) |D|^s \right) dt \right|^2 \ll (1 + |s(s-1)|^2) hY^{2\sigma} X^2 (\log^+ \frac{Xh}{Y})^2$$

Proof. This essentially follows by partial summation from Lemma 4.3.

Lemma 4.5 *Let a, h be integers with $(ah, 2N) = 1$. With notation as above and $\sigma \geq 0$, we have*

$$\sum_{\substack{n \leq X \\ hn_2 \neq b^2}} \left| \frac{1}{Y} \int_0^Y \sum_{\substack{0 < D \leq t \\ D \equiv a \,(\mathrm{mod}\, 8N)}} \left(\frac{D}{hn}\right) |D|^s dt \right|^2 \ll (1 + |s(s-1)|^2) hY^{2\sigma} X^2 (\log^+ \frac{Xh}{Y})^2.$$

Proof. The sum over D may be written as

$$\frac{1}{\phi(8N)} \sum_{\chi \bmod 8N} \overline{\chi}(a) \sum_{0 < D \leq t} \chi(D) \left(\frac{D}{hn}\right) |D|^s.$$

Applying the Cauchy-Schwarz inequality, each term in the sum over n is

$$\ll \frac{1}{\phi(8N)} \sum_{\chi} \sum_{\substack{n \leq X \\ hn_2 \neq b^2}} \left| \frac{1}{Y} \int_0^Y \sum_{0 < D \leq t} \chi(D) \left(\frac{D}{hn}\right) |D|^s dt \right|^2.$$

If $hn_2 \neq b^2$, then $\left(\frac{D}{hn_2}\right)$ is a nontrivial character of conductor prime to $8N$, and so $\chi\left(\frac{\cdot}{hn}\right)$ is a nontrivial character. Hence, the above summand is

$$\ll \frac{1}{\phi(8N)} \sum_{\chi} \sum_{n \leq X}{}^* \left| \frac{1}{Y} \int_0^Y \sum_{0 < D \leq t} \chi(D) \left(\frac{D}{hn}\right) |D|^s dt \right|^2$$

where the asterisk on the sum has the same meaning as earlier (that is, we range over those $n \leq X$ so that $\chi\left(\frac{\cdot}{hn}\right) \neq 1$.) Applying Lemma 4.4, we deduce the result.

Lemma 4.6 *Let a, h be integers with $(ah, 2N) = 1$ and $\beta = \pm 1$. With notation as above and $\sigma \geq 0$, we have*

$$\sum_{\substack{n \leq X \\ hn_2 \neq b^2}} \left| \frac{1}{Y} \int_0^Y \sum_{\substack{0 < \beta D \leq t \\ D \equiv a \,(\mathrm{mod}\, 8N)}} \left(\frac{D}{hn}\right) |D|^s dt \right|^2 \ll (1 + |s(s-1)|^2) h Y^{2\sigma} X^2 (\log^+ \frac{Xh}{Y})^2.$$

Proof. This follows immediately by noting that the previous result holds even if we sum over $0 < -D \leq t$. (This is clear since we can factor out $\chi(-1)\left(\frac{-1}{hn}\right)$ from the sum over D.)

§5 Estimates for some weighted sums

We need to record some lemmas that are minor variants of estimates that appear in [MM]. We collect them here.

Lemma 5.1 *Let $\beta = \pm 1$. For $\mathrm{Re}\, s_0 = \sigma_0$ and $\mathrm{Re}\, s_1 = \sigma_1$, $0 \leq \sigma_1 < 1/2$, $\sigma_0 > 0$, $a \equiv 1(\mathrm{mod}\, 4)$, and $(ah, 2N) = 1$, we have*

$$\sum_{\substack{n \leq U, n_2 h \neq b^2}} \frac{a(n)}{n^{1+s_1}} \chi_0^{(j)}(nh) \tilde{g}_Y^\beta(nh, 2s_0; a)$$

$$\ll (|s_0| + 1) h^{1/2} U^{1/2 - \sigma_1} Y^{1/2 + 2\sigma_0} \log Uh.$$

If $\sigma_1 > \frac{1}{2}$, we have

$$\sum_{\substack{n > U, n_2 h \neq b^2}} \frac{a(n)}{n^{1+s_1}} \chi_0^{(j)}(nh) \tilde{g}_Y^\beta(nh, 2s_0; a)$$

$$\ll (|s_0| + 1) h^{1/2} U^{1/2 - \sigma_1} Y^{1/2 + 2\sigma_0} \log Uh.$$

Proof. This is essentially Lemma 4 and Lemma 8 of [MM]. For example, to prove the first estimate, one checks that

$$\sum_{\substack{n \leq U \\ n_2 h \neq b^2}} \frac{a(n)}{n^{1+s_1}} \chi_0^{(j)}(nh) \sum_{\substack{0 < \beta m \leq t \\ m \equiv a (\mathrm{mod}\, 8N)}} \left(\frac{m}{nh}\right) m^{2s_0}$$

is

$$\ll (|s_0| + 1) h^{\frac{1}{2}} U^{\frac{1}{2}-\sigma_1} t^{\frac{1}{2}+2\sigma_0} (\log Uh).$$

Integrating over t then gives the stated result.

Lemma 5.2 Let $\beta = \pm 1$. If $\mathrm{Re}\, s_0 = \sigma_0$, and $\mathrm{Re}\, s_1 = \sigma_1$, $0 \leq \sigma_1 < 1/2$, $\sigma_0 > 0$, $a \equiv 1 (\mathrm{mod}\, 4)$, $(ah, 2N) = 1$, then

$$\sum_{n \leq U, n_2 h \neq b^2} \frac{a(n)}{n^{1+s_1}} g_Y^\beta (nh, 2s_0; a) \ll (|s_0| + 1) h^{1/2} U^{1/2-\sigma_1} Y^{1/2+2\sigma_0} (\log Y)(\log Uh).$$

If $\sigma_1 > \frac{1}{2}$, then

$$\sum_{n > U, n_2 h \neq b^2} \frac{a(n)}{n^{1+s_1}} g_Y^\beta (nh, 2s_0; a) \ll (|s_0| + 1) h^{1/2} U^{1/2-\sigma_1} Y^{1/2+2\sigma_0} (\log Y)(\log Uh).$$

Proof. This is essentially Lemma 5 and Lemma 9 of [MM].

Lemma 5.3 *We have*

$$\sum_{n \leq x} \frac{|a(n)|}{\sqrt{n}} \ll x (\log x)^{-\rho}$$

for some $\rho > 0$.

Proof. This is due to Rankin [Ra] (see [MM, Lemma 17] or Theorem IV.9.1).

Lemma 5.4. *For $0 \leq \sigma_1 < 1$, we have*

$$\frac{1}{Y} \sum_{n_2 h \neq b^2} \frac{a(n)}{n^{1+s_1}} \left(\int_1^Y \sum_{\substack{|D| \leq t \\ D \equiv a (\mathrm{mod}\, 8N)}} |D|^{2s_0} \left(\frac{D}{nh}\right) \, dt \right) \exp(-n/X)$$

$$\ll (|s_0| + 1) h^{\frac{1}{2}} Y^{2\sigma_0} \frac{X^{1-\sigma_1}}{1-\sigma_1} (\log X)^{-\rho}.$$

Proof. Use Lemma 2.2 and Lemma 5.3.

§6 The statements $A^{\pm}(\alpha)$ and $C^{\pm}(\alpha)$

Let $\beta = \pm 1$. In §3, we had defined $M_\beta(s_0, s_1, j, h)$ and $N_\beta(s_0, s_1, j, h)$. Consider the following statement:

$$A^\beta(\alpha): \qquad \frac{1}{Y} \int_1^Y \sum_{\substack{0 < \beta D \leq t \\ D \equiv a \,(\mathrm{mod}\, 8N)}} \left(\frac{D}{h}\right) |D|^{2s_0} L_{Dj^2}(1 + s_1, f)\, dt$$

$$= M_\beta(s_0, s_1, j, h) + N_\beta(s_0, s_1, j, h)$$

$$+ \mathbf{O}\left(Y^{1-\sigma_1 + 2\sigma_0} \left\{ \sqrt{h}(1 + |s_0| + |s_0 - s_1|)^2 \right.\right.$$

$$\left.\left. \times \,(\log Yh)^\alpha \log(h \log Y) + \sigma_{-3/4}(j)^3 \right\}\right)$$

for $a \equiv 1 \,(\mathrm{mod}\, 4)$, $(ahj, 2N) = 1$, $\sigma_0 \geq 0$ and $0 \leq \sigma_1 < \frac{1}{2}$.

Lemma 6.1 *Let* $\beta = \pm 1$. *Then* $A^\beta(2)$ *holds.*

Proof. Let $a \equiv 1 \,(\mathrm{mod}\, 4)$ and $(a, 2N) = 1$. Suppose also that $\sigma_0 \geq 0$ and that $0 \leq \sigma_1 < \frac{1}{2}$. Let $X \geq Y$ and consider the integral

$$\frac{1}{Y} \int_1^Y \sum_{\substack{0 < \beta D \leq t \\ D \equiv a \,(\mathrm{mod}\, 8N)}} \left(\frac{D}{h}\right) |D|^{2s_0} \left(\frac{1}{2\pi i} \int_{(2)} L_{Dj^2}(1 + s_1 + w, f) X^w \Gamma(w)\, dw\right) dt.$$

On the one hand, it is

$$\sum_{(n,j)=1} \frac{a(n)}{n^{1+s_1}} \tilde{g}_Y^\beta(nh, 2s_0; a) \, \exp(-n/X).$$

(Recall that we are writing

$$\tilde{g}_Y^\beta(n, 2s_0; a) = \frac{1}{Y} \int_1^Y \sum_{\substack{0 < \beta D \leq t \\ D \equiv a \,(\mathrm{mod}\, 8N)}} |D|^{2s_0} \left(\frac{D}{n}\right) dt$$

as usual.) Let us estimate the contribution of those n for which $n_2 h$ is not a square. Applying the Cauchy-Schwarz inequality, we see that the terms with $n \leq X$ are

$$\ll \left(\sum_{n \leq X} \frac{|a(n)|^2}{n^{\frac{5}{4}+\sigma_1}} \exp(-n/X)\right)^{\frac{1}{2}} \times \left(\sum_{\substack{n \leq X \\ n_2 h \neq b^2}} \frac{1}{n^{\frac{3}{4}+\sigma_1}} \left|\tilde{g}_Y^\beta(nh, 2s_0; a)\right|^2 \exp(-n/X)\right)^{\frac{1}{2}}.$$

We apply Rankin's estimate for the first factor. We find that it is

$$\ll X^{\frac{1}{2}(\frac{3}{4} - \sigma_1)}.$$

For the second factor, we apply Lemma 4.6, and find that it is

$$\ll (1+|s_0(2s_0-1)|)h^{\frac{1}{2}}Y^{2\sigma_0}X^{\frac{1}{2}(\frac{5}{4}-\sigma_1)}\log^+\frac{Xh}{Y}.$$

Putting these estimates together, we get an estimate of

$$\ll E_1 \overset{\text{def}}{=} (1+|s_0(2s_0-1)|)h^{\frac{1}{2}}Y^{2\sigma_0}X^{1-\sigma_1}\log^+\frac{Xh}{Y}$$

Similarly, the sum over terms $n > X$ is majorized by

$$\left(\sum_{n>X}\frac{|a(n)|^2}{n^{2\sigma_1}}\exp\left(-n/X\right)\right)^{\frac{1}{2}}\left(\sum_{\substack{n>X\\n_2h\neq b^2}}\frac{1}{n^2}|\tilde{g}_Y^\beta(nh,2s_0;a)|^2\exp\left(-n/X\right)\right)^{\frac{1}{2}}$$

and this is $\ll E_1$ also. If we choose $X \leq Y(\log Y)^\gamma$, we see that

$$E_1 \ll (1+|s_0(2s_0-1)|)h^{\frac{1}{2}}Y^{1+2\sigma_0-\sigma_1}(\log Yh)^{\gamma(1-\sigma_1)}\log(h\log Y).$$

The sum corresponding to the squares is equal to

$$\frac{1}{Y}\sum_{\substack{n_2h=b^2\\(n,j)=1}}\frac{a(n_1n_2)}{(n_1n_2)^{1+s_1}}\left(\frac{a}{n_1}\right)\int_1^Y\left(\sum_{\substack{0<\beta D\leq t\\D\equiv a(\mathrm{mod}\,8N)\\(D,2Nn_2h)=1}}|D|^{2s_0}\right)dt\,\exp(-n/X).$$

By Lemma 3.1, this is

$$= M_\beta(s_0,s_1,j,h) + \mathbf{O}(\sigma_{-3/4}(j)^3X^{-\sigma_1-\frac{1}{8}}Y^{1+2\sigma_0}).$$

On the other hand, moving the line of integration to the line $Re(w) = -\eta$, we get a residue at $w = 0$ equal to

$$\frac{1}{Y}\int_1^Y\sum_{\substack{0<\beta D\leq t\\D\equiv a(\mathrm{mod}\,8N)}}\left(\frac{D}{h}\right)|D|^{2s_0}L_{Dj^2}(1+s_1,f)dt.$$

plus an integral along the line $Re(w) = -\eta$. We proceed as in §3 and rewrite it as

$$\beta\omega\left(\frac{a}{N}\right)A^{-2s_1}\epsilon(a)\sum_{\substack{\delta^2\leq Y\\(\delta,2Nh)=1\\j|\delta}}\delta^{4s_0}\bar{\epsilon}(\delta^2)\sum_{d|\delta^2}\frac{\tilde{\mu}(d,f)}{d^{1+s_1}}\frac{1}{2\pi i}\int_{(-\eta)}\sum\frac{\bar{a}(n)}{n^{1-s_1-w}}$$

$$\times g_{Y/\delta^2}(ndh,2(s_0-s_1-w);a\bar{\delta}^2)\frac{\Gamma(1-s_1-w)}{\Gamma(1+s_1+w)}(X/A^2d)^w\Gamma(w)dw.$$

To estimate this integral, we split the sum over n into those for which $n_2 dh$ is a square, and the remaining values. If $\frac{1}{8} \leq \sigma_1 < \frac{1}{2}$, then by Lemma 3.2, we see that the first set contribute an amount which is

$$\ll Y^{1+2\sigma_0} X^{-\sigma_1}.$$

If $0 \leq \sigma_1 < \frac{1}{8}$, then it is

$$N_\beta(s_0, s_1, j, h) + \mathbf{O}(Y^{\frac{5}{8}+2\sigma_0} X^{\frac{3}{16}-\sigma_1}).$$

Using Lemma 5.2, we shall estimate the contribution of the second set. Split the sum over n at an intermediate point U to be specified later. For the terms with $n < U$, we move the line of integration to a line $Re(w) = -\eta_1$, with

$$\sigma_1 \leq \eta_1 < \sigma_1 + \frac{1}{2},$$

and for those with $n > U$, we move the line of integration to a line $Re(w) = -\eta_2$, with

$$2 > \eta_2 > 1.$$

Let us set $Z = Y/\delta^2$. The terms with $n < U$ then contribute an amount

$$\sum_{\delta^2 \leq Y} \delta^{4\sigma_0} \sum_{d|\delta^2} \frac{d^{\frac{1}{2}} \mathbf{d}(d)}{d^{1+\sigma_1}} \left(\frac{X}{A^2 d}\right)^{-\eta_1} |\Gamma(-\eta_1)|(1 + |s_0 - s_1 + \eta_1|)$$

$$\times (dh)^{\frac{1}{2}} U^{\frac{1}{2}-(\eta_1-\sigma_1)} Z^{\frac{1}{2}+2(\sigma_0-\sigma_1+\eta_1)} (\log Z)(\log U dh).$$

Now, if we choose $U = Y^2/X$, then

$$X^{-\eta_1} U^{\frac{1}{2}-\eta_1+\sigma_1} Y^{\frac{1}{2}+2\sigma_0-2\sigma_1+2\eta_1} = Y^{3/2+2\sigma_0} X^{-\frac{1}{2}-\sigma_1}.$$

We choose

$$\eta_1 = \sigma_1 + \frac{1}{2} - \nu$$

for a ν satisfying

$$\frac{1}{4} < \nu < \min(\frac{1}{2}, \sigma_1 + \frac{3}{8})$$

say. (This choice ensures that $\eta_1 > 1/8$ and in particular is bounded away from zero.) Then the sum over δ above is

$$\sum_{\delta^2 \leq Y} \delta^{-1+4(\nu-\frac{1}{2})} \sum_{d|\delta^2} d^{\frac{1}{2}-\nu} \mathbf{d}(d)$$

and this is

$$\ll \sum_{\delta^2 \leq Y} \delta^{4\nu-3}(\delta^2)^{\frac{1}{2}-\nu}\mathbf{d}(\delta^2)^2$$

$$\ll \sum_{\delta^2 \leq Y} \delta^{2\nu-2}\mathbf{d}(\delta^2)^2 \ll 1.$$

Hence the above is

$$\ll h^{\frac{1}{2}}Y^{3/2+2\sigma_0}X^{-\frac{1}{2}-\sigma_1}(\log Y)(\log Yh)(1+|s_0-s_1|).$$

The contribution of the terms with $n > U$ is entirely similar to the above with η_2 replacing η_1. In moving to the line $-\eta_2$, we also encounter a residue at $w = -1$. Using Lemma 5.2, this residue is easily shown to be

$$\ll (1+|s_0-s_1|)\sqrt{h}X^{-\frac{1}{2}-\sigma_1}Y^{3/2+2\sigma_0}(\log Y)(\log Yh).$$

Now we choose (say) $\eta_2 = 3/2$. Then we see the total contribution is

$$\ll (1+|s_0-s_1|)h^{\frac{1}{2}}Y^{3/2+2\sigma_0}X^{-\frac{1}{2}-\sigma_1}(\log Y)(\log Yh)$$

and for $X \geq Y$ this is

$$\ll (1+|s_0-s_1|)h^{\frac{1}{2}}Y^{1+2\sigma_0-\sigma_1}(\log Y)(\log Yh).$$

This proves $A^{\beta}(2)$.

To proceed further, we introduce the following smoothing operator. We define

$$I_U^0(f) = f, \quad I_U(f) = I_U^1(f) = \frac{1}{U}\int_U^{2U} f(u)du.$$

Moreover, for $n \geq 1$, we set

$$I_U^n(f) = \frac{1}{U}\int_U^{2U} I_t^{n-1}(f)dt.$$

Let us also set

$$A(u) = \sum_{\substack{n<u \\ n_2 h \neq b^2 \\ (n_2,j)=1}} \frac{a(n)}{n^{1+s_1}}\tilde{g}_Y^{\beta}(nh,2s_0;a)$$

and for $\sigma_1 > \frac{1}{2}$,

$$B(u) = \sum_{\substack{n>u \\ n_2 h \neq b^2 \\ (n_2,j)=1}} \frac{a(n)}{n^{1+s_1}}\tilde{g}_Y^{\beta}(nh,2s_0;a).$$

Lemma 6.2 *Assume $A^\beta(\alpha)$ and suppose $\sigma_0 \geq 0$. If $0 \leq \sigma_1 < \frac{1}{2}$, then we have for $U \geq Y$*

$$I_U^4(A) \ll Y^{1-\sigma_1+2\sigma_0}\left\{\sqrt{h}(1+|s_0|+|s_0-s_1|)^2(\log Yh)^\alpha \log(h\log Y) + \sigma_{-3/4}(j)^3\right\}.$$

Moreover, if $1 > \sigma_1 > \frac{1}{2}$ and $U \geq Y$, then we have

$$I_U^3(B) \ll Y^{1+2\sigma_0}U^{-\sigma_1}\left\{\sqrt{h}(1+|s_0|+|s_0-s_1|)^2(\log Yh)^\alpha \log(h\log Y) + \sigma_{-3/4}(j)^3\right\}.$$

Proof. Let us set

$$A^*(u) \;=\; \sum_{\substack{n<u \\ (n_2,j)=1}} \frac{a(n)}{n^{1+s_1}}\tilde{g}_Y^\beta(nh, 2s_0; a).$$

Then

$$I_U^4(A^*(u))$$

$$= \frac{1}{2\pi i}\int_{(c)}\left\{\frac{1}{Y}\int_1^Y \sum_{\substack{0<\beta D\leq t \\ D\equiv a(\mathrm{mod}\,8N)}}\left(\frac{D}{h}\right)|D|^{2s_0}L_{Dj^2}(1+s_1+w, f)dt\right\}I_U^4(u^w)\frac{dw}{w}$$

provided $\sigma_1 + c > \frac{1}{2}$. We move the line of integration to $\mathrm{Re}(w) = c_1 < 0$ where c_1 is chosen so that $0 < \sigma_1 + c_1 < \frac{1}{2}$. Appealing to $A^\beta(\alpha)$, this is equal to

$$M_\beta(s_0,s_1,j,h) \;+\; N_\beta(s_0,s_1,j,h)$$

$$+ \frac{1}{2\pi i}\int_{(c_1)}(M_\beta(s_0,s_1+w,j,h)+N_\beta(s_0,s_1+w,j,h))I_U^4(u^w)\frac{dw}{w}$$

$$+ \mathbf{O}\left(Y^{1-\sigma_1+2\sigma_0}\left\{\sqrt{h}(1+|s_0|+|s_0-s_1|)^2(\log Yh)^\alpha \log(h\log Y)+\sigma_{-3/4}(j)^3\right\}\right).$$

The condition $U \geq Y$ is used in obtaining the **O**-term. Using the definition of N_β and Lemma 3.3, it is easy to check that

$$N_\beta(s_0,s_1,j,h) \;\ll\; Y^{\frac{1}{2}+2\sigma_0}(\log Y)(|s_1|+1)^{1-2\sigma_1}\mathbf{d}(h_0)h_0^{\sigma_1-\frac{1}{2}}$$

for $0 \leq \sigma_1 < \frac{1}{2}$ and $\sigma_0 \geq 0$. (Here, $h = h_0 h_1^2$ with h_0 squarefree.) The same estimate holds for the integral above involving N_β. Next, we observe that

$$M_\beta(s_0,s_1,j,h) \;+\; \frac{1}{2\pi i}\int_{(c_1)}M_\beta(s_0,s_1+w,j,h)I_U^4(u^w)\frac{dw}{w}$$

gives the contribution of squares in $I_U^4(A^*(u))$ (that is, those n for which $n_2h = b^2$). This proves the first statement of the lemma. For the second part, consider the sum

$$S \;=\; I_U^3\Big(\sum_{\substack{n>u \\ (n_2,j)=1}} \frac{a(n)}{n^{1+s_1}}\tilde{g}_Y^\beta(nh, 2s_0; a)\Big).$$

We shall estimate it using partial summation. Set

$$C(x) \;=\; \sum_{\substack{u<n<x \\ (n_2,j)=1}} \frac{a(n)}{n^{\frac12+s_1}} \tilde{g}_Y^\beta(nh, 2s_0; a).$$

We have

$$S \;=\; I_U^3\Big(\int_u^\infty t^{-\frac12} dC(t)\Big) \;=\; I_U^3\Big(\frac12\int_u^\infty C(t) t^{-3/2} dt\Big)$$

and this is equal to

$$\frac{1}{2\pi i}\int_{(c)}\left(\frac{1}{Y}\int_1^Y \sum_{\substack{0<\beta D\le t \\ D\equiv a(\mathrm{mod}\,8N)}} \left(\frac{D}{h}\right)|D|^{2s_0} L_{Dj^2}\Big(\frac12+s_1+w, f\Big) dt\right) I_U^3(u^{w-\frac12})\frac{2dw}{1-2w}$$

provided $\sigma_1+c>1$ and $0<c<\frac12$. Now move the line of integration to $\mathrm{Re}(w)=c_1$ where $\frac12 \le \sigma_1 + c_1 < 1$. (This is possible because $\frac12 < \sigma_1 < 1$.) Now apply $A^\beta(\alpha)$ on the line c_1. The "main term" in $A^\beta(\alpha)$ gives

$$\frac{1}{2\pi i}\int_{(c_1)}\Big\{M_\beta\Big(s_0, s_1+w-\frac12\Big) + N_\beta\Big(s_0, s_1+w-\frac12\Big)\Big\}I_U^3(u^{w-\frac12})\frac{2dw}{1-2w}.$$

The M_β term gives the square terms in S. The N_β term is zero if $\sigma_1 > 5/8$. If $\sigma_1 \le 5/8$ then it is

$$\ll (|s_1|+1)^{1-2(\sigma_1+c_1-\frac12)}U^{c_1-\frac12}Y^{\frac12+2\sigma_0}h_0^{c_1+\sigma_1-\frac12}\mathbf{d}(h_0).$$

The **O** term in $A^\beta(\alpha)$ gives an amount which is

$$Y^{1-(\sigma_1+c_1-\frac12)+2\sigma_0}U^{c_1-\frac12}\{\sqrt{h}(1+|s_0|+|s_0-s_1|)^2$$
$$\times (\log Yh)^\alpha \log(h\log Y) + \sigma_{-3/4}(j)^3\}.$$

Choosing $c_1 = \frac12 - \sigma_1$ gives the result.

Next we consider the following statement which provides estimates for functions similar to $A(u)$ and $B(u)$ in which we restrict the sum to fundamental discriminants.

$$C^\beta(\alpha) : I_X^4\Big(\sum_{\substack{n\le u \\ n_2h\ne b^2 \\ (n,j)=1}} \frac{a(n)}{n^{1+s_1}} g_Y^\beta(nh, 2s_0; a)\Big)$$

$$\ll Y^{\frac12+2\sigma_0}X^{\frac12-\sigma_1}$$

$$\frac{1}{2\sigma_1-1}\Big\{\sqrt{h}(1+|s_0|+|s_0-s_1|)^2(\log Yh)^\alpha \log(h\log Y) + \sigma_{-3/4}(j)^3\Big\}$$

for $a \equiv 1 \pmod 4$, $(ah, 2N) = 1$, $(2Nh, j) = 1$ $\sigma_0 \geq 0$ and $0 < \sigma_1 < \frac{1}{2}$. Moreover, for $X \geq Y$

$$I_X^3 \left(\sum_{\substack{n > u \\ n_2 h \neq b^2 \\ (n_2, j) = 1}} \frac{a(n)}{n^{1+s_1}} g_Y^\beta(nh, 2s_0; a) \right)$$

$$\ll Y^{\frac{1}{2}+2\sigma_0} X^{\frac{1}{2}-\sigma_1} \{\sqrt{h}(1 + |s_0| + |s_0 - s_1|)^2 (\log Yh)^\alpha (\log(h \log Y)) + \sigma_{-3/4}(j)^3\}$$

for $a \equiv 1 \pmod 4$, $(ah, 2N) = 1$, $\sigma_0 \geq 0$ and $1 > \sigma_1 > \frac{1}{2}$.

In the next result, we exhibit the relationship between $A^\beta(\alpha)$ and $C^\beta(\alpha)$.

Proposition 6.3 *If $A^\beta(\alpha)$ holds, then $C^\beta(\alpha)$ holds.*

Proof. For the first half of $C^\beta(\alpha)$, we have

$$I_X^4 \left(\sum_{\substack{n \leq u \\ n_2 h \neq b^2 \\ (n, j) = 1}} \frac{a(n)}{n^{1+s_1}} g_Y^\beta(nh, 2s_0; a) \right) du$$

$$= I_X^4 \left(\sum_{\substack{n \leq u \\ n_2 h \neq b^2 \\ (n, j) = 1}} \frac{a(n)}{n^{1+s_1}} \frac{1}{Y} \int_1^Y \sum_{\substack{0 < \beta D \leq t \\ D \equiv a \pmod{8N}}} \left(\frac{D}{nh} \right) |D|^{2s_0} \sum_{g^2 | D} \mu(g) dt \right)$$

$$= \sum_{\substack{g^2 \leq Y \\ (g, 2Nh) = 1}} \mu(g) g^{4s_0} I_X^4 \left(\sum_{\substack{n \leq u \\ n_2 h \neq b^2 \\ (n_2, jg) = 1}} \frac{a(n)}{n^{1+s_1}} \tilde{g}_{Y/g^2}^\beta(nh, 2s_0; a\bar{g}^2) \right).$$

To estimate this, we split the sum over g into two parts at

$$V = \sqrt{\frac{Y}{X}}.$$

Lemma 4.6, Rankin's estimate and the Cauchy-Schwarz inequality imply that the first part is

$$\ll \sum_{g \leq V} g^{4\sigma_0} \left(\frac{Y}{g^2} \right)^{2\sigma_0} X^{1-\sigma_1} \sqrt{h}(1 + |s_0(2s_0 - 1)|) \log^+(Xhg^2/Y)$$

and this is

$$\ll \sqrt{h}(1 + |s_0(2s_0 - 1)|) Y^{2\sigma_0} X^{1-\sigma_1} \sqrt{\frac{Y}{X}} (\log h)$$

$$\ll \sqrt{h}(1 + |s_0(2s_0 - 1)|) Y^{\frac{1}{2}+2\sigma_0} X^{\frac{1}{2}-\sigma_1} (\log h).$$

For the second part, Lemma 6.2 implies that it is

$$\ll \sum_{\sqrt{Y}>g>V} g^{4\sigma_0}\left(\frac{Y}{g^2}\right)^{1-\sigma_1+2\sigma_0}$$

$$\left\{(1+|s_0|+|s_0-s_1|)^2\sqrt{h}(\log Yh)^{\alpha}(\log(h\log Y)) + \sigma_{-3/4}(j)^3\right\}$$

$$\ll Y^{1-\sigma_1+2\sigma_0}(1-2\sigma_1)^{-1}\left(\sqrt{\frac{Y}{X}}\right)^{2\sigma_1-1}$$

$$\left\{(1+|s_0|+|s_0-s_1|)^2\sqrt{h}(\log Yh)^{\alpha}\log(h\log Y) + \sigma_{-3/4}(j)^3\right\}$$

and this is

$$\ll Y^{\frac{1}{2}+2\sigma_0}X^{\frac{1}{2}-\sigma_1}(1-2\sigma_1)^{-1}$$

$$\left((1+|s_0|+|s_0-s_1|)^2\sqrt{h}(\log Yh)^{\alpha}\log(h\log Y) + \sigma_{-3/4}(j)^3\right).$$

This proves the first part of $C^{\beta}(\alpha)$.

For the second half, we can apply Lemma 6.2 directly to the entire range of the sum over g.

Proposition 6.4 *If $C^{\beta}(\alpha)$ holds then $A^{\beta}(\frac{4}{5}\alpha)$ holds.*

Proof. We proceed exactly as in the proof of Lemma 6.1, only in place of Lemma 5.2, we use $C^{\beta}(\alpha)$ to estimate the contribution of the second set. For each fixed value of δ, we will split the sum over n at an intermediate point u_δ to be specified later. For the terms with $n < u_\delta$, we move the line of integration to a line $Re(w) = -\eta_1$, and bounded away from zero, with

$$\sigma_1 < \eta_1 < \sigma_1 + \frac{1}{2}.$$

For those with $n > u_\delta$, we move the line of integration to a line $Re(w) = -\eta_2$, with

$$\begin{cases} \sigma_1 + \frac{1}{2} < \eta_2 < 1 & \text{if } \sigma_1 < 1/8 \\ 1 < \eta_2 < 1+\sigma_1 & \text{if } 1/8 < \sigma_1 < \frac{1}{2}. \end{cases}$$

When $\eta_2 > 1$, we also pick up a residue R from the pole at $w = -1$. (We remark that the point 1/8 was only used as a convenient breaking point. The essential difference between the two cases is whether σ_1 is bounded away from zero or not.) Let us set $Z = Y/\delta^2$ and

$$U = U_\delta = Z(\log(Zh+100))^{\gamma}$$

for some $2 > \gamma > 0$ to be specified later. By the first part of $C^\beta(\alpha)$ and the mean value theorem, there exists $V = V_\delta$ in the interval $(U_\delta, 2U_\delta)$ such that

$$I_V^3 \left(\sum_{\substack{n < u \\ n_2 h \neq b^2 \\ (n_2, j) = 1}} \frac{\bar{a}(n)}{n^{1 - s_1 - w}} g_Z^\beta(ndh, 2(s_0 - s_1 - w); a\bar{\delta}^2)) \right.$$

$$\ll Z^{\frac{1}{2} + 2(\sigma_0 - \sigma_1 + \eta_1)} U_\delta^{\frac{1}{2} - (\eta_1 - \sigma_1)}$$

$$\times \frac{1}{(2(\eta_1 - \sigma_1) - 1)} \left(\sqrt{dh}(1 + |s_0 - s_1 - w| + |s_0|)^2 \right.$$

$$\times (\log Y dh)^\alpha \log(dh \log Y) + \sigma_{-3/4}(j)^3)$$

Moreover, as $V \geq Z$

$$I_V^3 \left(\sum_{\substack{n > u \\ n_2 h \neq b^2 \\ (n_2, j) = 1}} \frac{\bar{a}(n)}{n^{1 - s_1 - w}} g_Z^\beta(ndh, 2(s_0 - s_1 - w); a\bar{\delta}^2)) \right.$$

$$\ll Z^{\frac{1}{2} + 2(\sigma_0 - \sigma_1 + \eta_2)} U_\delta^{\frac{1}{2} - (\eta_2 - \sigma_1)}$$

$$\times \left(\sqrt{dh}(1 + |s_0 - s_1 - w| + |s_0|)^2 (\log Y dh)^\alpha \log(dh \log Y) + \sigma_{-3/4}(j)^3 \right).$$

For each δ we apply the operator

$$I_{V_\delta}^3$$

with respect to the variable u_δ to both sides of the basic equation. Then, the left hand side and the residue do not change (as they are independent of u_δ). And the $n < u_\delta$ summed over all δ contribute an amount

$$\sum_{\delta^2 \leq Y} \delta^{4\sigma_0} \sum_{d | \delta^2} \frac{d^{\frac{1}{2}} \mathbf{d}(d)}{d^{1 + \sigma_1}} \log d |\Gamma(-\eta_1)| (X/d)^{-\eta_1} Z^{\frac{1}{2} + 2(\sigma_0 - \sigma_1 + \eta_1)} U_\delta^{\frac{1}{2} - (-\sigma_1 + \eta_1)}$$

$$\frac{1}{(\eta_1 - \sigma_1 - \frac{1}{2})} \left\{ \sqrt{dh}(|s_0| + |s_0 - s_1| + 1)^2 (\log Z dh)^\alpha \log(h \log Y) + \sigma_{-3/4}(j)^3 \right\}$$

$$\ll Y^{\frac{1}{2} + 2(\sigma_0 - \sigma_1 + \eta_1)} X^{-\eta_1} (Y(\log Y h)^\gamma)^{\frac{1}{2} + \sigma_1 - \eta_1}$$

$$\times \frac{1}{(\eta_1 - \sigma_1 - \frac{1}{2})} \sum_{\delta^2 \leq Y} \delta^{-2 + 2\sigma_1 - 2\eta_1} \sum_{d | \delta^2} \frac{d^{\frac{1}{2} + \eta_1} \mathbf{d}(d)}{d^{1 + \sigma_1}}$$

$$\left((|s_0| + |s_0 - s_1| + 1)^2 \sqrt{dh}(\log Y h)^\alpha \log(h \log Y) + \sigma_{-3/4}(j)^3 \right)$$

The above may be simplified to

$$\ll Y^{1 + 2\sigma_0 - \sigma_1 + \eta_1} X^{-\eta_1} (\log Y h)^{\gamma(\frac{1}{2} + \sigma_1 - \eta_1)} \frac{1}{(\eta_1 - \sigma_1 - \frac{1}{2})}$$

$$\left((|s_0| + |s_0 - s_1 + \eta_1| + 1)^2 \sqrt{h}(\log Y h)^\alpha \log(h \log Y) + \sigma_{-3/4}(j)^3 \right)$$

$$\sum_{\delta^2 \leq Y} \delta^{-2 + 2\sigma_1 - 2\eta_1} \sum_{d | \delta^2} \mathbf{d}(d) d^{\eta_1 - \sigma_1} \log d.$$

This is seen to be

$$\ll E_2 \overset{\text{def}}{=} Y^{1+2\sigma_0-\sigma_1} \left(\frac{Y}{X}\right)^{\eta_1} (\log Yh)^{\gamma(\frac{1}{2}+\sigma_1-\eta_1)}$$

$$\frac{1}{(\eta_1-\sigma_1-\frac{1}{2})} \left((|s_0|+|s_0-s_1+\eta_1|+1)^2 \sqrt{h}(\log Yh)^{\alpha}\log(h\log Y) + \sigma_{-3/4}(j)^3 \right)$$

Now, using the second part of $C^{\beta}(\alpha)$, we see that the sum over δ of the terms with $n > u_{\delta}$ contribute an amount

$$\ll E_3 \overset{\text{def}}{=} Y^{1+2\sigma_0-\sigma_1} \left(\frac{Y}{X}\right)^{\eta_2} (\log Yh)^{\gamma(\frac{1}{2}+\sigma_1-\eta_2)}$$

$$\left\{ \sqrt{h}(|s_0|+|s_0-s_1+\eta_2|+1)^2 (\log Yh)^{\alpha} \log(h\log Y) + \sigma_{-3/4}(j)^3 \right\}.$$

We now choose X so that

$$Y^{2\sigma_0} X^{1-\sigma_1} = Y^{1+2\sigma_0-\sigma_1} \left(\frac{Y}{X}\right)^{\eta_1} (\log Yh)^{\alpha+\gamma(\frac{1}{2}+\sigma_1-\eta_1)}.$$

This is ensured if

$$\gamma = \frac{\alpha}{\frac{1}{2}+2\eta_1-2\sigma_1}$$

and

$$X = Y(\log Yh)^{\gamma}.$$

With this choice, (and with E_1 as in Lemma 6.1),

$$E_1 + E_2 \ll Y^{1+2\sigma_0-\sigma_1} (\log Yh)^{\frac{\alpha(1-\sigma_1)}{\frac{1}{2}+2\eta_1-2\sigma_1}}$$

$$\frac{1}{(\eta_1 - \sigma_1 - \frac{1}{2})} \left\{ (|s_0|+|s_0-s_1+\eta_1|+1)^2 \sqrt{h} \log(h\log Y) + \sigma_{-3/4}(j)^3 \right\}.$$

Simplifying, we see that if $\eta_1 = \sigma_1 + \frac{1}{2} - \nu$, with $\nu = 1/8$ (say) then the exponent of $\log Yh$ is

$$(1-\sigma_1)(\frac{\alpha}{\frac{1}{2}+2\eta_1-2\sigma_1}) = (1-\sigma_1)\frac{\alpha}{\frac{3}{2}-2\nu}$$

and this is

$$\le \frac{4}{5}\alpha.$$

Thus,

$$E_1 + E_2 \ll Y^{1+2\sigma_0-\sigma_1} (\log Yh)^{\frac{4}{5}\alpha}$$

$$\{(|s_0|+|s_0-s_1|+1)^2 h^{\frac{1}{2}} (\log(h\log Y)) + \sigma_{-3/4}(j)^3 \}.$$

The error term E_3 gives a similar quantity with the exponent of $\log Yh$ equal to

$$\alpha - \left(2\eta_2 - \sigma_1 - \frac{1}{2}\right)\frac{\alpha}{2\eta_1 - 2\sigma_1 + \frac{1}{2}}.$$

We choose

$$\eta_2 = \begin{cases} \sigma_1 + 5/8 & \text{if } \sigma_1 < 1/8 \\ 17/16 & \text{if } 1/8 < \sigma_1 < 1/2. \end{cases}$$

Then, the above is

$$\leq \frac{4}{5}\alpha.$$

Hence,

$$E_3 \ll Y^{1+2\sigma_0-\sigma_1}\{(|s_0| + |s_0 - s_1| + 1)^2 h^{\frac{1}{2}}(\log Yh)^{\frac{4}{5}\alpha}\log(h\log Y) + \sigma_{-3/4}(j)^3\}.$$

The second part of $C^\beta(\alpha)$ easily implies a similar estimate holds for R. We conclude that

$$E_1 + E_2 + E_3 + R$$
$$\ll Y^{1+2\sigma_0-\sigma_1}\{(|s_0| + |s_0 - s_1| + 1)^2 h^{\frac{1}{2}}$$
$$(\log Yh)^{\frac{4}{5}\alpha}\log(h\log Y) + \sigma_{-3/4}(j)^3\}.$$

and so $A^\beta(\frac{4}{5}\alpha)$ holds.

§7 Proof of main result

As a consequence of the results of the previous sections, we deduce the following crucial result.

Theorem 7.1. *Let $\beta = \pm 1$, and $\lambda > 0$. Then $C^\beta(\lambda)$ holds.*

Proof. From Lemma 6.1, we know that $A^\beta(2)$ is true. From Proposition 6.3, we deduce that $C^\beta(2)$ holds. Suppose we have established $C^\beta(\alpha)$. By Proposition 6.4, we deduce that $A^\beta(\frac{4\alpha}{5})$ is true. By Proposition 6.3, we deduce that $C^\beta(\frac{4\alpha}{5})$ holds. Iterating this proves the result.

Remark. We see also that given $\lambda > 0$, the statement $A^\beta(\lambda)$ holds.

Proof of Theorem 1.2. Let $a \equiv 1 \pmod 4$, $(a, 2N) = 1$. Consider the integral

$$\frac{1}{Y}\int_1^Y \sum_{\substack{|D|\leq t \\ D\equiv a(\bmod 8N)}} \frac{1}{2\pi i}\int_{(2)} L_D(1 + w, f)X^w\Gamma(w)dwdt.$$

On the one hand, it is

$$\frac{1}{Y} \sum \frac{a(n)}{n} \left(\int_1^Y \sum_{\substack{|D| \le t \\ D \equiv a \,(\mathrm{mod}\, 8N)}} \left(\frac{D}{n}\right) dt \right) \exp(-n/X).$$

Let us estimate the contribution of those n for which n_2 is not a square. We apply Lemma 5.4 and find that the sum is

$$\ll X (\log X)^{-\rho}.$$

Thus, if we choose $X \le Y (\log Y)^\nu$ with $0 < \nu < \rho$, we see that this is

$$\ll Y (\log Y)^{-\kappa}$$

for some $\kappa > 0$. Now, the contribution of those values of n for which n_2 is a square is seen to be equal to

$$\sum_{n_2 = b^2} \frac{a(n)}{n} \left(\frac{a}{n_1}\right) \left(\frac{1}{Y} \int_1^Y \sum_{\substack{|D| \le t \\ D \equiv a \,(\mathrm{mod}\, 8N) \\ (D,n)=1}} 1 \, dt \right) \exp(-n/X)$$

and from Lemma 3.5, we know that this is

$$= C(0,0,1,1)Y + \mathbf{O}(YX^{-1/8})$$

where we recall that

$$C(0,0,1,1) = \frac{1}{8N} \sum \frac{a(n_1 n_2^2)}{n_1 n_2^2} \left(\frac{a}{n_1}\right) \frac{\phi(n_2)}{n_2}.$$

(Note that we are summing over both positive and negative values of D here.) On the other hand, moving the line of integration to the line $\mathrm{Re}(w) = -\eta$, $0 < \eta < 1$, we get a residue at $w = 0$ equal to

$$\frac{1}{Y} \int_1^Y \sum_{\substack{|D| \le t \\ D \equiv a \,(\mathrm{mod}\, 8N)}} L_D(1,f) dt$$

and an integral

$$\omega\left(\frac{a}{N}\right) \epsilon(a) \sum_{\substack{\delta^2 \le Y \\ (\delta, 2N)=1}} \bar{\epsilon}(\delta^2) \sum_{d|\delta^2} \frac{\tilde{\mu}(d,f)}{d} \frac{1}{2\pi i} \int_{(-\eta)} \sum \frac{\bar{a}(n)}{n^{1-w}} \times$$

$$\times \left(\frac{\delta^2}{Y} \int_1^{Y/\delta^2} \sum_{\substack{|D_0| \le t \\ D_0 \delta^2 \equiv a \,(\mathrm{mod}\, 8N)}} |D_0|^{-2w} (\mathrm{sgn} D_0) \left(\frac{D_0}{nd}\right) dt \right) \times \qquad (*)$$

$$\times \frac{\Gamma(1-w)}{\Gamma(1+w)} (X/A^2 d)^w \Gamma(w) dw$$

which has been simplified in the (by now) familiar way. Let us write it as

$$\Sigma_1 + \Sigma_2$$

where in Σ_2 we only include terms with $n_2 d = b^2$ and Σ_1 contains the remaining terms.

First, consider the contribution of the terms with $n_2 d = b^2$. We have

$$\sum_{\substack{|D_0| \le t/\delta^2 \\ D_0 \delta^2 \equiv a \,(\mathrm{mod}\, 8N) \\ (D_0, n_2 d) = 1}} |D_0|^{-2w} (\mathrm{sgn} D_0)$$

$$= \frac{1}{\phi(8N)} \sum_{\psi} \bar{\psi}(a\bar{\delta}^2) \sum_{m \le \sqrt{t}/\delta} \mu(m) \chi_0^{(n_2 d)}(m^2) \psi(m^2) m^{-4w}$$

$$\times \sum_{|h| \le t/\delta^2 m^2} \chi_0^{(n_2 d)}(h) \psi(h) |h|^{-2w} (\mathrm{sgn}\ h)$$

and it is not hard to check that the above is

$$\ll \mathbf{d}(8N n_2 d) \left(\frac{t}{\delta^2} \right)^{2\eta + \frac{1}{2}}.$$

Inserting this into the big expression above, we see that

$$\Sigma_2 \ll \sum_{\delta^2 \le Y} \sum_{d | \delta^2} \frac{d^{\frac{1}{2}} \mathbf{d}(d)}{d} \left(\frac{X}{d} \right)^{-\eta} |\Gamma(-\eta)| \left(\frac{Y}{\delta^2} \right)^{2\eta + \frac{1}{2}} \mathbf{d}(8N d).$$

Choosing $\eta = 1/4$ for example, we see that

$$\Sigma_2 \ll Y X^{-1/4}.$$

We now estimate Σ_1. By $C^\beta(\lambda)$ we know that for $0 < \eta_1 < \frac{1}{2}$ and $\mathrm{Re}(w) = -\eta_1$, there exists $U \in (X, 2X)$ such that

$$I_U^3 \left(\frac{\delta^2}{Y} \int_1^{Y/\delta^2} \sum_{\substack{n \le u \\ n_2 d \neq b^2}} \frac{\bar{a}(n)}{n^{1-w}} f_t^\beta(nd, -2w; a\bar{\delta}^2) dt \right)$$

$$\ll (2\eta_1 - 1)^{-1} d^{\frac{1}{2}} (|w| + 1)^2 (Y/\delta^2)^{\frac{1}{2} + 2\eta_1} X^{\frac{1}{2} - \eta_1}$$

$$\left((\log Y d/\delta^2)^\lambda (\log d \log Y/\delta^2) \right).$$

As $d \le \delta^2 \le Y$ we see that the above is

$$\ll d^{\frac{1}{2}} (|w| + 1)^2 (Y/\delta^2)^{\frac{1}{2} + 2\eta_1} X^{\frac{1}{2} - \eta_1} (\log Y)^\lambda \log(d \log Y).$$

Also, for $\frac{1}{2} < \eta_2 < 1$, and $\mathrm{Re}(w) = -\eta_2$, $2X \geq U \geq X \geq Y$,

$$I_U^3 \left(\frac{\delta^2}{Y} \int_1^{Y/\delta^2} \sum_{\substack{n > u \\ n_2 d \neq b^2}} \frac{\bar{a}(n)}{n^{1-w}} f_t^\beta(nd, -2w; a\bar{\delta}^2) dt \right)$$
$$\ll d^{\frac{1}{2}}(|w| + 1)^2 (Y/\delta^2)^{\frac{1}{2} + 2\eta_2} X^{\frac{1}{2} - \eta_2} (\log Y)^\lambda \log(d \log Y).$$

We shall choose η_1 and η_2 bounded away from 0 and 1 respectively. Using the estimate

$$|\tilde{\mu}(d, f)| \ll \mathbf{d}(d) d^{\frac{1}{2}}$$

we deduce that

$$\Sigma_1 \ll \sum_{i=1}^2 Y^{\frac{1}{2} + 2\eta_i} X^{\frac{1}{2} - 2\eta_i} (\log Y)^\lambda (\log \log Y) \times \sum_{\delta^2 \leq Y} \frac{1}{\delta^{1 + 4\eta_i}} \sum_{d | \delta^2} \frac{d^{\frac{1}{2}} \mathbf{d}(d)}{d} d^{\eta_i} d^{\frac{1}{2}} \log d$$

and this is

$$\ll X^{\frac{1}{2}} Y^{\frac{1}{2}} (\log Y)^\lambda (\log \log Y) \sum_\delta \frac{\mathbf{d}(\delta^2)^2}{\delta} \log \delta \left\{ \left(\frac{Y}{X\delta} \right)^{2\eta_1} + \left(\frac{Y}{X\delta} \right)^{2\eta_2} \right\}.$$

Simplifying, we see that if we set

$$X = Y(\log Y)^\nu$$

then

$$\Sigma_1 \ll Y(\log Y)^{\nu/2 + \lambda - 2\eta_1 \nu}(\log \log Y).$$

Now, if we choose $\lambda = \nu/10$ for any $0 < \nu < \rho$, and $\eta_1 = 2/5$(say) then

$$\Sigma_1 \ll Y(\log Y)^{-\nu/5}(\log \log Y).$$

Together with our earlier estimate of Σ_2, this proves the main theorem.

Exercises

1. Deduce Theorem 1.1 from Theorem 1.2 by showing that if there are only a finite number of fundamental discriminants D with $L_D(1, f) \neq 0$ then

$$\sum_{\substack{|D| \leq Y \\ D \equiv a \,(\mathrm{mod}\, 8N)}} L_D(1, f) \ll \sqrt{Y}(\log Y).$$

2.   For a prime p not dividing N, let us define

$$B_p(s) = \sum_{j=0}^{\infty} \frac{a(p^{2j})}{p^{2js}} \quad \text{and} \quad C_p(s) = \sum_{j=0}^{\infty} \frac{a(p^{2j+1})}{p^{(2j+1)s}}.$$

Then, we have the identities

$$B_p(s) = \left(1 - \frac{\alpha_p^2}{p^{2s}}\right)^{-1} \left(1 - \frac{\beta_p^2}{p^{2s}}\right)^{-1} \left(1 - \frac{\epsilon(p)p}{p^{2s}}\right)^{-1} \left(1 - \frac{1}{p^{4s-2}}\right)$$

and

$$C_p(s) = p^{-s}(1 + \epsilon(p)p) \left(1 - \frac{\alpha_p^2}{p^{2s}}\right)^{-1} \left(1 - \frac{\beta_p^2}{p^{2s}}\right)^{-1}$$

where $a(p) = \alpha_p + \beta_p$ and $|\alpha_p| = |\beta_p| = p^{\frac{1}{2}}$.

3.   For d not dividing N, define the function

$$F_d(s, f) = \sum_{n=1}^{\infty} a(dn^2)(dn^2)^{-s}.$$

Then,

$$F_{d_0}(s, f) = \left(\sum_{n=1}^{\infty} \frac{a(n^2)}{n^{2s}}\right) \prod_{p \mid d_0} B_p(s)^{-1} C_p(s).$$

Using the identity

$$\sum_{n=1}^{\infty} \frac{a(n^2)}{n^{2s}} = \prod_p B_p(s) = L(2s, Sym^2)\zeta(4s - 2)^{-1},$$

deduce that

$$F_d(s, f) = L(2s, Sym^2)\zeta(4s - 2)^{-1}$$
$$\prod_{p \mid d} \left(1 - \frac{\epsilon(p)p}{p^{2s}}\right) \left(1 - \frac{1}{p^{4s-2}}\right)^{-1} \left(\frac{1 + \epsilon(p)p}{p^s}\right).$$

All of these formal manipulations are valid for $\text{Re}(s) > 1$.

4.　For $(d, 2ND) = 1$, consider

$$\tilde{F}_d(s, f, D) = \sum_{\substack{m_2 d = b^2 \\ (m, D) = 1}} \frac{a(m)}{m^{1+s}} \left(\frac{a}{m_1} \right) \frac{\phi(m_2 d)}{m_2 d}.$$

Show that $\tilde{F}_d(s, f, D)$ is defined for $\mathrm{Re}(s) \geq 0$ and has an analytic continuation for $\mathrm{Re}(s) > -1/4$. Moreover, for $\mathrm{Re}(s) \geq -3/16$ (say), it satisfies the estimate of Lemma 3.3: in the notation of that lemma,

$$\tilde{F}_d(s, f, D) \ll c_1^{\nu(d_0)} (d_0)^{-\sigma - \frac{1}{2}} |L(2s + 2, Sym^2) \zeta(2 + 4s)^{-1}| \prod_{p \mid D} (1 + \frac{1}{p^{1+2\sigma}})^3.$$

5.　Prove that if f has trivial character,

$$L(1, f) = 2 \sum_{n=1}^{\infty} \frac{a(n)}{n} e^{-2\pi n / \sqrt{N}}.$$

Deduce that

$$\sum_{|D| \leq Y} L_D(1, f) \ll Y$$

where the sum ranges over D which are prime to N. State and prove a similar result for forms f with non-trivial character.

6.　Let

$$A(X, \chi) = \sum_{n=1}^{\infty} a(n) \chi(n) n^{-1} e^{-2\pi n / X}.$$

Using the large sieve inequality, prove the estimate

$$\sum_{D \leq Y} \sum_{\chi \bmod D} {}^* |A(X, \chi)|^4 \ll (X + Y)^{2+\epsilon}.$$

Here, the sum over χ ranges over primitive characters. Deduce the estimate of Iwaniec [I]

$$\sum_{|D| \leq Y} |L_D(1, f)|^4 \ll Y^{2+\epsilon}.$$

*7.　Prove the asymptotic formulae for averages of higher derivatives:

$$\frac{1}{Y} \int_1^Y \sum_{|D| \leq t} L_D^{(j)}(1, f) dt \sim c_j Y (\log Y)^j$$

for $j \geq 0$ and constants c_j.

References

[BC] P. T. Bateman and S. Chowla, Averages of character sums, *Proc. Amer. Math. Soc.*, **1** (1950), 781–787.

[FH] S. Friedberg and J. Hoffstein, Non-vanishing theorems for automorphic L-functions on $GL(2)$, *Annals of Math.*, **142** (1995), 385–423.

[FS] A. S. Fainleib and O. Saparnijazov, Dispersion of real character sums and the moments of $L(1, \chi)$, (Russian) *Izv. Akad. Nauk. USSR, Ser. Fiz.-Mat. Nauk*, **19** (1975), 24–29.

[I] H. Iwaniec, On the order of vanishing of modular L-functions at the critical point, Sém. de Théorie des Nombres Bordeaux, **2** (1990), 365–376.

[J] M. Jutila, On character sums and class numbers, *J. Number Theory*, **5** (1973), 203–214.

[MV] H.L. Montgomery and R.C. Vaughan, Mean values of character sums, Canadian J. Math., **31** (1979), 476–487.

[M] V. Kumar Murty, A non-vanishing theorem for quadratic twists of modular L-functions, preprint, 1991.

[MM] M. Ram Murty and V. Kumar Murty, Mean values of derivatives of modular L-series, *Annals of Math.*, **133** (1991), 447–475.

[MS] V. Kumar Murty and T. Stefanicki, Non-vanishing of quadratic twists of L-functions attached to automorphic representations of GL(2) over **Q**, preprint, 1994.

[Ra] R. Rankin, Sums of powers of cusp form coefficients II, *Math. Ann.*, **272** (1985), 593–600.

[Ro] D. Rohrlich, Non-vanishing of L-functions for GL_2, *Invent. Math.*, **97** (1989), 381–403.

[Sh] G. Shimura, On the periods of modular forms, *Math. Ann.*, **229** (1977), 211–221.

[W1] J. Waldspurger, Sur les valeurs de certaines fonctions L automorphe en leur centre de symétrie, *Comp. Math.*, **54** (1985), 173–242.

[W2] J. Waldspurger, Correspondances de Shimura et quaternions, *Forum Math.*, **3** (1991), 219–307.

Chapter 7
Selberg's Conjectures

§1 Selberg's class of Dirichlet series

In a fundamental paper [S], Selberg defined a general class of Dirichlet series and formulated basic conjectures concerning them. Selberg's conjectures concern Dirichlet series, which admit analytic continuations, Euler products and functional equations.

The Riemann zeta function is the simplest example of a function in the family S of functions $F(s)$ of a complex variable s satisfying the following properties:

(i) (Dirichlet series) For $\mathrm{Re}(s) > 1$,

$$F(s) = \sum_{n=1}^{\infty} \frac{a_n}{n^s},$$

where $a_1 = 1$ and we will write $a_n(F) = a_n$ for the coefficients of the Dirichlet series;

(ii) (Analytic continuation) $F(s)$ extends to a meromorphic function so that for some integer $m \geq 0$, $(s-1)^m F(s)$ is an entire function of finite order;

(iii) (Functional equation) There are numbers $Q > 0$, $\alpha_i > 0$, $\mathrm{Re}(r_i) \geq 0$, so that

$$\Phi(s) = Q^s \prod_{i=1}^{d} \Gamma(\alpha_i s + r_i) F(s)$$

satisfies

$$\Phi(s) = w \overline{\Phi(1 - \bar{s})}$$

for some complex number w with $|w| = 1$;

(iv) (Euler product)

$$F(s) = \prod_{p} F_p(s)$$

where

$$F_p(s) = \exp\left(\sum_{k=1}^{\infty} \frac{b_{p^k}}{p^{ks}}\right)$$

where $b_{p^k} = \mathbf{O}(p^{k\theta})$ for some $\theta < 1/2$, where p runs over prime numbers.

(v)   (Ramanujan hypothesis) $a_n = \mathbf{O}(n^\epsilon)$ for any fixed $\epsilon > 0$.

Note that the family S is multiplicatively closed, and so is a multiplicative monoid.

All known examples of elements in S are automorphic L-functions. In all of these cases, $F_p(s)$ is an inverse of a polynomial in p^{-s} of bounded degree.

Selberg [S] introduced this family to study the value distribution of finite linear combinations of Dirichlet series with Euler products and functional equations. For this purpose, he introduced the important concept of a primitive function and made significant conjectures about them.

A function $F \in S$ is called **primitive** if the equation $F = F_1 F_2$ with $F_1, F_2 \in S$ implies $F = F_1$ or $F = F_2$. As we shall see below, one of the most serious consequences of the Selberg conjectures is that S has unique factorization into primitive elements. It is not difficult to show that every element of S can be factored into primitive elements. This is a consequence of an old theorem of Bochner [B], though Selberg [S] and more recently Conrey and Ghosh [CG] seem to have found it independently.

Selberg conjectures:

Conjecture A: For all $F \in S$, there exists a positive integer n_F such that

$$\sum_{p \leq x} \frac{|a_p(F)|^2}{p} = n_F \log\log x + \mathbf{O}(1).$$

In Proposition 2.5, we shall describe n_F more explicitly.

Conjecture B:
(i)   for any primitive function F, $n_F = 1$ so that

$$\sum_{p \leq x} \frac{|a_p(F)|^2}{p} = \log\log x + \mathbf{O}(1);$$

(ii)   for two distinct primitive functions F and F',

$$\sum_{p \leq x} \frac{a_p(F)\overline{a_p(F')}}{p} = \mathbf{O}(1).$$

Thus, in some sense, the primitive functions form an orthonormal system.

In his paper [S], Selberg investigates the consequences of his conjectures to the value distribution of $\log F(\sigma + it)$ for $\sigma = 1/2$ or σ very near to $1/2$. Selberg also conjectures the analogue of the Riemann hypothesis for the functions $F \in \mathcal{S}$.

It is not difficult to see that Conjecture B implies Conjecture A. By Proposition 2.4 below, Conjecture B also implies that the factorization into primitives in $\mathcal{S}$ is unique. It seems central, therefore, to classify the primitive functions.

To this end, it is natural to define the **dimension** of F as

$$\dim F = 2\alpha_F$$

where

$$\alpha_F = \sum_{i=1}^{d} \alpha_i.$$

By Proposition 2.2 below, this concept is well-defined. Selberg conjectures that the dimension of F is always a non-negative integer. This question was previously raised by Vignéras [V].

Bochner's work can be used to classify primitive functions of dimension one in the case $\alpha_1 = 1/2$. They are (after a suitable translation) the classical zeta function of Riemann and the classical L-functions of Dirichlet. If the α_i are rational numbers, then this is also a complete list of (primitive) functions of dimension one (see [Mu]).

We will show that Conjecture B implies Artin's conjecture concerning the holomorphy of non-abelian L-series attached to irreducible Galois representations. More precisely, let k be an algebraic number field and K/k a finite Galois extension with group G. Let ρ be an irreducible representation on the n-dimensional complex vector space V. As explained in Chapter 2, for each prime ideal $\mathfrak{p}$ of k, let $V_{\mathfrak{p}}$ be the subspace of V fixed by the inertia group $I_{\mathfrak{p}}$ of $\mathfrak{p}$. Set

$$L_{\mathfrak{p}}(s, \rho) = \det(1 - \rho(\sigma_{\mathfrak{p}})\mathbf{N}\mathfrak{p}^{-s}|V_{\mathfrak{p}})^{-1}$$

where $\sigma_{\mathfrak{p}}$ is the Frobenius automorphism of the prime ideal $\mathfrak{p}$ of k, $\mathbf{N}$ is the absolute norm from k to $\mathbb{Q}$. Define

$$L(s, \rho; K/k) = \prod_{\mathfrak{p}} L_{\mathfrak{p}}(s, \rho).$$

(Sometimes, we write $L(s, \rho)$ if the field extension is clear. Since $L(s, \rho)$ depends only on the character χ of ρ, we will also sometimes write $L(s, \chi)$ or $L(s, \chi, K/k)$ for $L(s, \rho, K/k)$.) Clearly, the Artin L-function $L(s, \rho)$ is a product of L-functions attached to irreducible constituents of ρ. We have:

Artin's conjecture: If ρ is irreducible $\neq 1$, then $L(s, \rho, K/k)$ extends to an entire function of s.

§2 Basic consequences

We record in this section the results of Bochner [B], Selberg [S] and Conrey-Ghosh [CG].

Proposition 2.1 (Bochner [B]) *If $F \in \mathcal{S}$, and $\alpha_F > 0$, then $\alpha_F \geq 1/2$.*

Remark. In their paper, Conrey and Ghosh [CG] give a simple proof of this and also treat the case $\alpha_F = 0$. They prove that the constraint $b_n = \mathbf{O}(n^\theta)$ for some $\theta < 1/2$ implies there is no element in $\mathcal{S}$ with $\alpha_F = 0$ except the constant function 1.

Proposition 2.2 (Selberg [S]) *Let $N_F(T)$ be the number of zeroes $\rho = \beta + i\gamma$ of $F(s)$ satisfying $0 < \gamma \leq T$. Then,*

$$N_F(T) = \frac{\alpha_F}{\pi} T(\log T + c) + S_F(T) + \mathbf{O}(1),$$

where c is a constant and $S_F(T) = \mathbf{O}(\log T)$.

If $F = F_1 F_2$, then clearly $N_F(T) = N_{F_1}(T) + N_{F_2}(T)$ so that $\alpha_F = \alpha_{F_1} + \alpha_{F_2}$. Thus, if F is such that $\alpha_F < 1$, then F is necessarily primitive. The following is now immediate.

Proposition 2.3 (Conrey-Ghosh [CG]) *Every $F \in \mathcal{S}$ has a factorization into primitive functions.*

Proof. If F is not primitive, then $F = F_1 F_2$ and by the above, $\alpha_F = \alpha_{F_1} + \alpha_{F_2}$. By Proposition 2.1, each of α_{F_1}, α_{F_2} is strictly less than α_F. Continuing this process, we find that the process terminates because Proposition 2.1 implies the number of factors is $\leq 2\alpha_F$. This completes the proof.

We see immediately that the Riemann zeta function and the classical Dirichlet functions $L(s, \chi)$ with χ a primitive character are primitive in the sense of Selberg. Indeed, the Γ-factor appearing in the functional equation is $\Gamma(s/2)$ or $\Gamma((s+1)/2)$ and the result is now clear from Proposition 2.1.

Conjecture B forces the factorization in Proposition 2.3 to be unique. Indeed, suppose that F had two factorizations into primitive functions:

$$F = F_1 \ldots F_r = G_1 \ldots G_t$$

where $F_1, \ldots, F_r, G_1, \ldots, G_t$ are primitive functions. Without loss of generality, we may suppose that no G_i is an F_1. But then,

$$a_p(F_1) + \cdots + a_p(F_r) = a_p(G_1) + \cdots + a_p(G_t)$$

so that

$$\sum_{p \leq x} \frac{\overline{a_p(F_1)}(a_p(F_1) + \cdots + a_p(F_r))}{p} = \sum_{p \leq x} \frac{\overline{a_p(F_1)}(a_p(G_1) + \cdots + a_p(G_t))}{p}.$$

As $x \to \infty$, the left hand side tends to infinity, whereas the right hand side is bounded since no G_i is an F_1. This contradiction proves:

Proposition 2.4 (Conrey-Ghosh [CG]) *Conjecture B implies that every $F \in S$ has a unique factorization into primitive functions.*

In the next proposition, we describe n_F.

Proposition 2.5
(a) *If $F \in S$ and $F = F_1^{e_1} \cdots F_r^{e_r}$ is a factorization into primitive functions, then Conjecture B implies*

$$n_F = e_1^2 + \cdots + e_r^2.$$

(b) *Conjecture B implies that F is primitive if and only if $n_F = 1$.*

Proof. We have

$$a_p(F) = \sum_{i=1}^{r} e_i a_p(F_i)$$

and so computing the asymptotic behaviour of

$$\sum_{p \leq x} \frac{|a_p(F)|^2}{p}$$

and using Conjecture B yields the result.

§3 Artin's conjecture and Selberg's conjectures

We now discuss Artin's conjecture in the context of Selberg's conjectures. We begin by showing that Selberg's conjectures imply the holomorphy of non-abelian L-functions. Let χ be the character of the representation ρ. We will write $L(s, \chi, K/k)$ for $L(s, \rho, K/k)$.

Theorem 3.1 *Conjecture B implies Artin's conjecture.*

Proof. We adhere to the notation introduced above. Let $\tilde{K}$ be the normal closure of K over $\mathbb{Q}$. Then, $\tilde{K}/k$ is Galois, as well as $\tilde{K}/\mathbb{Q}$, and χ can be thought of as a character $\tilde{\chi}$ of $\mathrm{Gal}(\tilde{K}/k)$. By the property of Artin L-functions (see [A]),

$$L(s, \tilde{\chi}, \tilde{K}/k) = L(s, \chi, K/k).$$

Moreover, if $\mathrm{Ind}\,\tilde{\chi}$ denotes the induction of $\tilde{\chi}$ from $\mathrm{Gal}(\tilde{K}/k)$ to $\mathrm{Gal}(\tilde{K}/\mathbb{Q})$, then

$$L(s, \tilde{\chi}, \tilde{K}/k) = L(s, \mathrm{Ind}\,\tilde{\chi}, \tilde{K}/\mathbb{Q}),$$

by the invariance of Artin L-functions under induction. Hence, we can write

$$L(s, \chi, K/k) = \prod_{\phi} L(s, \phi, \tilde{K}/\mathbb{Q})^{m(\phi)}$$

where the product is over irreducible characters ϕ of $\mathrm{Gal}(\tilde{K}/\mathbb{Q})$ and $m(\phi)$ are non-negative integers. To prove Artin's conjecture, it suffices to show that $L(s, \phi, \tilde{K}/\mathbb{Q})$ is entire for each irreducible character ϕ of $\mathrm{Gal}(\tilde{K}/\mathbb{Q})$. By Brauer's induction theorem and the Artin reciprocity law, we can write

$$L(s, \phi, \tilde{K}/\mathbb{Q}) = \frac{L(s, \chi_1)}{L(s, \chi_2)}$$

where χ_1 and χ_2 are characters of $\mathrm{Gal}(\tilde{K}/\mathbb{Q})$ and $L(s, \chi_1)$, $L(s, \chi_2)$ are entire functions, being products of Hecke L-functions. Thus, they belong to $\mathcal{S}$ and hence, by Proposition 2.4, have a unique factorization into primitive functions. We can therefore write

$$L(s, \phi) = \prod_{i=1}^{m} F_i(s)^{e_i}, \qquad e_i \in \mathbb{Z}.$$

By comparing the p-th Dirichlet coefficient of both sides, we get

$$\phi(p) = \sum_{i=1}^{m} e_i a_p(F_i)$$

from which we obtain

$$\sum_{p \le x} \frac{|\phi(p)|^2}{p} = \sum_{p \le x} \frac{1}{p} \left| \sum_{i=1}^{m} e_i a_p(F_i) \right|^2.$$

Conjecture B gives the asymptotic behaviour of the right hand side:

$$\sum_{p \le x} \frac{|\phi(p)|^2}{p} = \left(\sum_{i=1}^{m} e_i^2 \right) \log \log x + \mathbf{O}(1).$$

Decompose the sum on the left hand side according to the conjugacy class C of $\mathrm{Gal}(\tilde{K}/\mathbb{Q})$ to which the Frobenius automorphism σ_p belongs:

$$\sum_{p \le x} \frac{|\phi(p)|^2}{p} = \sum_{C} |\phi(g_C)|^2 \sum_{\substack{p \le x \\ \sigma_p \in C}} \frac{1}{p},$$

where g_C is any element of C. By the Chebotarev density theorem

$$\sum_{\substack{p \le x \\ \sigma_p \in C}} \frac{1}{p} = \frac{|C|}{|G|} \log \log x + \mathbf{O}(1).$$

Hence,

$$\sum_{p \le x} \frac{|\phi(p)|^2}{p} = \sum_{C} \frac{|C|}{|G|} |\phi(g_C)|^2 \log \log x + \mathbf{O}(1).$$

But ϕ is irreducible and so,

$$\sum_C \frac{|C|}{|G|}|\phi(g_C)|^2 = (\phi, \phi) = 1.$$

Therefore, the left hand side is

$$\log \log x + \mathbf{O}(1)$$

as $x \to \infty$. We deduce that

$$\sum_{i=1}^{m} e_i^2 = 1$$

from which follows $m = 1$ and $e_1 = \pm 1$. Thus, $L(s, \phi) = F(s)$ or $1/F(s)$, where $F(s)$ is primitive and analytic everywhere except possibly at $s = 1$. However, $L(s, \phi)$ has trivial zeroes and so the latter possibility cannot arise. We conclude that $L(s, \phi) = F(s)$ is primitive and entire.

Corollary 3.2   *Let $K/\mathbb{Q}$ be Galois and let χ be an irreducible character of* $\mathrm{Gal}(K/\mathbb{Q})$. *Conjecture B implies that $L(s, \chi)$ is primitive.*

Proof. This is evident from the last line in the proof of the previous lemma. Or we can derive it as follows. By the previous theorem, $F = L(s, \chi) \in \mathcal{S}$ and by the Chebotarev density theorem, $n_F = 1$. The result now follows from Proposition 2.5 (b).

Of course, Dedekind's conjecture that the zeta function of a number field is always divisible by the Riemann zeta function follows from Artin's conjecture. However, it is rather interesting to note that the unique factorisation conjecture is sufficient to deduce this. Indeed, if K is a number field, $\tilde{K}$ its Galois closure, and $\zeta_K(s)$ is the Dedekind zeta function of K, then

$$\zeta_{\tilde{K}}(s)/\zeta_K(s) = F(s)$$

is entire by the Aramata-Brauer theorem. By the same theorem,

$$\zeta_{\tilde{K}}(s)/\zeta(s) = G(s)$$

is also entire. Since $\zeta(s)$ is primitive, it appears as a primitive factor in $\zeta_{\tilde{K}}(s) = \zeta(s)G(s)$. Since $\zeta_{\tilde{K}}(s) = \zeta_K(s)F(s)$ and F is entire, $\zeta(s)$ must appear in the unique factorization of $\zeta_K(s)$. This is Dedekind's conjecture.

The Selberg conjectures refer to the analytic behaviour of Dirichlet series at the edge of the critical strip. (There are other conjectures relating special values of Dirichlet series inside the critical strip, namely the Deligne conjectures and the Birch-Swinnerton-Dyer conjectures to cite specific instances.) A consequence of

Conjecture B is that if F is any primitive function which is not the Riemann zeta function, then

$$\sum_{p \leq x} \frac{a_p(F)}{p^{1+it}} = \mathbf{O}(1).$$

In particular, no primitive function should vanish on $\sigma = 1$. Thus, Selberg's conjectures imply that no element of $\mathcal{S}$ vanishes on $\mathrm{Re}(s) = 1$. Most likely, the Selberg class consists only of automorphic L-functions in which case there is a general non-vanishing result of Jacquet and Shalika [JS].

Many of our interesting consequences, notably the Artin conjecture, utilised the unique factorization conjecture. Perhaps this can be attacked by other means. Indeed, given r distinct primitive functions $F_1, \ldots, F_r$, one would expect the existence of complex numbers $s_1, \ldots, s_r$ such that $F_i(s_j) = 0$ if and only if $i = j$. If this were the case, then clearly, the unique factorization conjecture is true.

The classification of primitive functions is a fundamental problem. From the work of Bochner and Vignéras, it follows that if F has dimension 1 and all the α_i are rational numbers, then $d = 1$ and $\alpha_1 = 1/2$. It then follows, essentially from the same works, that F must either be the Riemann zeta function or a purely imaginary translate of a classical Dirichlet L-function attached to a non-trivial primitive character. It is shown in [Mu] that if π is an irreducible cuspidal automorphic representation of $GL_2(\mathcal{A}_{\mathbb{Q}})$, then $L(s, \pi)$ is primitive if the Ramanujan conjecture is true. In particular, the L-function attached to a normalised holomorphic cuspidal Hecke eigenform is a primitive function which is in Selberg's class (by Deligne's theorem).

Exercises

1. If F and $G \in \mathcal{S}$, define

$$F \times G = \prod_p H_p(s) \quad \text{where} \quad H_p(s) = \exp\left(\sum_{k=1}^{\infty} k b_{p^k}(F) b_{p^k}(G) p^{-ks} \right).$$

Define $F \in \mathcal{S}$ to be *simple* if $F \times \overline{F}$ extends to an analytic function for $\mathrm{Re}(s) \geq 1/2$ except for a simple pole at $s = 1$. Show that if $F \in \mathcal{S}$ is simple and entire, then $F(1 + it) \neq 0$ for all $t \in \mathbb{R}$. (Hint: a simple function has at most a simple pole at $s = 1$.)

2. If $F \in \mathcal{S}$ is simple and $F \times F$ has analytic continuation to $\mathrm{Re}(s) = 1$, then show that $F(1 + it) \neq 0$ for all $t \in \mathbb{R}$.

3. Let $F \in \mathcal{S}$ and assume that $a_n(F) \geq 0$. If $F \neq 1$, show that Selberg's conjectures imply that $\zeta | F$.

4. Show that the Dedekind zeta function of $\mathbb{Q}(2^{1/3})$ factors into a product of two distinct primitive functions, one of which is the Riemann zeta function and the other of dimension 2.

5. If $F, G \in \mathcal{S}$ are such that $a_p(F) = a_p(G)$ and $a_{p^2}(F) = a_{p^2}(G)$ for all but finitely many primes p, show that $F = G$.

6. For $F \in \mathcal{S}$, denote by $Z_F(T)$ the multiset of zeros of $F(s)$ in the region $\mathrm{Re}(s) \geq 1/2$ and $|\mathrm{Im}(s)| \leq T$. For $F, G \in \mathcal{S}$, suppose the symmetric difference satisfies

$$|Z_F(T)\Delta Z_G(T)| = \mathbf{o}(T)$$

    as $T \to \infty$. Show that $F = G$.

Exercises 5 and 6 are from [MM].

References

[A]   E. Artin, Collected papers, Springer-Verlag, New York-Berlin, 1982.

[B]   S. Bochner, On Riemann's functional equation with multiple gamma factors, *Annals of Mathematics*, **67** (1958) 29–41.

[CG] B. Conrey and A. Ghosh, On the Selberg class of Dirichlet series, *Duke Math. Journal*, **72** No. 3, (1993) 673–693.

[JS] H. Jacquet and J.A. Shalika, A non-vanishing theorem for zeta functions of GL_n, *Inventiones Math.*, **38** (1976) p. 1–16.

[M]   M. Ram Murty, A motivated introduction to the Langlands program, in *Advances in Number Theory* (eds. F. Gouvea and N. Yui), pp. 37–66, Clarendon Press, Oxford, 1993.

[M1] M. Ram Murty, Selberg's conjectures and Artin L-functions, *Bulletin of the Amer. Math. Soc.*, **31** (1) (1994) p. 1–14.

[MM] M. Ram Murty and V. Kumar Murty, Strong multiplicity one for Selberg's class, *C.R. Acad. Sci. Paris*, **319** (Series I) (1994) p. 315–320.

[Mu] M. Ram Murty, Selberg conjectures and Artin L-functions, II, in *Current Trends in Mathematics and Physics, A tribute to Harish-Chandra*, (edited by S. D. Adhikari), Narosa Publishing House, 1995.

[S]   A. Selberg, Old and new conjectures and results about a class of Dirichlet series, Collected Papers, Volume II, pp. 47–63, Springer-Verlag.

[V]   M.F. Vignéras, Facteurs gamma et équations fonctionelles, *Lecture notes in mathematics*, **627** Springer-Verlag, Berlin-New York, 1976.

Chapter 8
Suggestions for Further Reading

In [Iw], Iwaniec considers a weighted sum

$$\sum_D \mu^2(D)L'_D(1,f)F(D/Y)$$

where F is a smooth function, compactly supported in $\mathbb{R}^+$ with positive mean value. He establishes an asymptotic formula for it of the form

$$\alpha Y \log Y + \beta Y + \mathbf{O}(Y^{\frac{13}{14}+\epsilon})$$

with some constants $\alpha \neq 0$ and β which depend on f and the test function F. From this, he is able to deduce that given any $\epsilon > 0$, and $Y > C(\epsilon)$, the number of fundamental discriminants of $D \leq Y$ such that $L'_D(1,f) \neq 0$ is at least $Y^{\frac{2}{3}-\epsilon}$. He is able to do this by establishing the upper bound

$$\sum_{d \leq Y} |L'_D(1,f)|^4 \ll Y^{2+\epsilon}$$

and then using the Cauchy-Schwarz inequality:

$$\left|\sum_D \mu^2(D)L'_D(1,f)F(D/Y)\right| \leq \#\{D \leq Y : L'_D(1,f) \neq 0\}^{3/4}\{\sum_D |L'_D(1,f)|^4\}^{1/4}.$$

The method is capable of generalization and extension. For instance, see Murty and Stefanicki [MS] and Stefanicki [St], as well as [PP].

In [GV], Goldfeld and Viola formulate the following conjecture. Let us suppose that we have a Dirichlet series

$$L_1(s) = \sum_{n=1}^{\infty} \frac{a_n}{n^s}$$

which converges absolutely in some half-plane. Define

$$L_2(s) = \sum_{n=1}^{\infty} \frac{a_{n^2}}{n^s}.$$

Let N are R be fixed integers. Let $(D, NR) = 1$, with D a fundamental discriminant. For any real character $\chi \bmod |D|$, we assume that $L_1(s, \chi)$ given by

$$L_1(s, \chi) = \sum_{n=1}^{\infty} \frac{a_n \chi(n)}{n^s}$$

extends to an entire function and satisfies a functional equation of the following type:

$$A_\chi^s T_\chi(s) L_1(s, \chi) = w_\chi A_\chi^{k-s} T_\chi(k - s) L_1(k - s, \bar{\chi})$$

where

$$A_\chi > 0, \quad k > 0, \quad w_\chi = w\epsilon(D)\chi(R)$$

with $|w| = 1$, and ϵ a primitive Dirichlet character $\bmod N$, and where

$$A_\chi^s T_\chi(s) L_1(s, \chi)$$

is an entire function of s.

Here $T_\chi(s)$ denotes a product of gamma factors

$$T_\chi(s) = \begin{cases} \prod_{i=1}^{J^+} \Gamma(s + \alpha_i^+) & \text{if } D > 0 \\ \prod_{i=1}^{J^-} \Gamma(s + \alpha_i^-) & \text{if } D < 0 \end{cases}$$

for positive integers J^+, J^-, and real numbers $\alpha_i^+, \alpha_i^- > -k/2$ depending only on $L_1(s)$. For convenience, we will write

$$T_\chi(s) = \prod_{i=1}^{J} \Gamma(s + \alpha_i)$$

it being clear that J and the α_i depend on the sign of D. We also assume that

$$A_\chi = f(|D|)$$

where $f(x)$ is a non-decreasing C^1 function of $x \geq 1$. We also suppose that $L_2(s)$ has an Euler product:

$$L_2(s) = \prod_{p} \prod_{i=1}^{2\lambda} \left(1 - \gamma_{p,i} p^{-s}\right)^{-\delta_i}$$

and a pole of order $\rho \geq 0$ at $s = k$. Under these conditions, Goldfeld and Viola conjecture that as $D \to \infty$,

$$\sideset{}{'}\sum_{|D| \leq X} L_1 \left(k/2, \left(\frac{D}{\cdot} \right) \right)$$

$$= (1 + \mathbf{o}(1)) \frac{1}{\rho!} [z^\rho L_2(k + 2z)]_{z=0} \sideset{}{'}\sum_{|D| \leq X} (1 + w_\chi) \log^\rho f(|D|) \prod_{p|D} \prod_{i=1}^{2\lambda} (1 - \gamma_{p,i} p^{-k})^{\delta_i},$$

where the dash on the sum means that we sum over fundamental discriminants. In Chapters 5 and 6, we have established special cases of this conjecture.

It is natural to consider this conjecture for the quadratic twists of a fixed automorphic L-function on GL_r. (See [Mu] for an introduction to terminology and notation.) It is then possible to prove upper bounds of general r. This has recently been done by Y. Zhang [Z, pp. 54–60]. She obtains that if π is an irreducible cuspidal automorphic representation of $GL_r(\mathcal{A}_\mathbb{Q})$, where $\mathcal{A}_\mathbb{Q}$ denotes the adele ring of the rational number field, and π has trivial central character, then

$$\sideset{}{'}\sum_{|D| \leq X} L(1/2, \pi \otimes \left(\frac{D}{\cdot} \right)) \ll X^{(r+1)/2} \log^2 X$$

for $r > 1$. In case $r = 1$ or $r = 2$, one can establish the Goldfeld-Viola conjecture. Indeed, in case $r = 1$, the method of Chapter 5 can be extended to grossenchar-acters and we leave it as an exercise for the reader. The case of $r = 2$ has been elucidated in Chapter 6 for holomorphic cuspidal automorphic representations. In the non-holomorphic case, this was dealt with in [MS].

The non-vanishing of L-functions at the center of the critical strip often seems to have arithmetic meaning as is seen by the Birch and Swinnerton-Dyer conjec-tures or more generally the conjectures of Deligne and Beillinson. To cite another instance, the famous theorem of Waldspurger shows that given a cuspidal automor-phic representation π of $PGL_2(\mathcal{A}_F)$, the corresponding representation under the Howe correspondence is an automorphic representation of the metaplectic cover of $SL_2(\mathcal{A}_F)$ if and only if there is a quadratic character χ such that $L(1/2, \pi \otimes \chi) \neq 0$. (See [PS].)

In some cases, the non-vanishing result of the L-function twisted by a Hecke character, not necessarily quadratic, is already enough for certain arithmetical applications as in the work of Ash and Ginzburg [AG].

The methods are equally adaptable for average values of quadratic twists of L-functions evaluated not at the center of the critical strip but at other points of the complex plane. One may also consider other twists and get analogous non-vanishing theorems. For instance, Barthel and Ramakrishnan [BR] have proved that given any irreducible, unitary, cuspidal automorphic representation π of GL_r over a field F, and any complex number s_0 with $\mathrm{Re}(s_0) \notin (1/(2r-2), 1 - (1/(2r-$

2))), there are infinitely many ray class characters χ of F such that $L(s_0, \pi \otimes \chi) \neq 0$. Non-vanishing near $\mathrm{Re}(s) = 1$ is the subject of investigation in [HR]. Non-vanishing at $s = 1/2$ would have consequences for the construction of p-adic L-functions associated to cuspidal automorphic representation of GL_{2r} as in the work of Ash and Ginzburg [AG].

Rohrlich [R1], proved the following non-vanishing theorem on GL_2. Let π be an irreducible cuspidal automorphic representation of GL_2 over any number field F and let s_0 be a complex number. Then, there are infinitely many ray class characters of F (of finite order) such that $L(s_0, \pi \otimes \chi) \neq 0$. Some applications are given in [R2]. Friedberg and Hoffstein [FH] give necessary and sufficient conditions for the existence of a quadratic ray class character with this property.

There are other related results such as the recent work of Luo, Rudnick and Sarnak [LRS] on the Selberg eigenvalue conjecture. This again is an extension of the methods outlined in the previous chapters.

Recent work of Böcherer, Furusawa and Schulze-Pillot [BFS] raises the question of the simultaneous non-vanishing of quadratic twists of two Hecke eigenforms.

There is the problem of Merel [Me] which asks for non-vanishing at $s = 1/2$ of the L-functions of the twists of a given eigenform by even Dirichlet characters. Such a result will have applications in determining good upper bounds for torsion of elliptic curves over cyclotomic fields.

References

[AG] A. Ash and D. Ginzburg, p-adic L-functions for $GL(2n)$, *Invent. Math.*, **116**(1994), 27–73.

[BFS] S. Böcherer, M. Furusawa and R. Schulze-Pillot, On Whittaker coefficients of some metaplectic forms, *Duke Math. J.*, **76**(1994), 761–772.

[BR] L. Barthel and D. Ramakrishnan, A nonvanishing result for twists of L-functions of $GL(n)$, *Duke Math. J.*, **74**(1994), 681–700.

[FH] S. Friedberg and J. Hoffstein, Non-vanishing theorems for automorphic L-functions on $GL(2)$, *Annals of Math.*, **142** (1995) pp. 385–423.

[GV] D. Goldfeld and C. Viola, Mean values of L-functions associated to elliptic, Fermat, and other curves at the center of the critical strip, *Journal of Number Theory*, **11** (1979) pp. 305–320.

[HR] J. Hoffstein and D. Ramakrishnan, Siegel zeros and cusp forms, *Int. Math. Res. Not.*, 1995, pp. 279–308.

[Iw] H. Iwaniec, On the order of vanishing of modular L-functions at the critical point, *Séminaire de Théorie des Nombres, Bordeaux*, **2** (1990) pp. 365–376.

[LRS] W. Luo, Z. Rudnick and P. Sarnak, On Selberg's eigenvalue conjecture, *Geom. and Func. Anal.*, **5**(1995), 387–401.

[Me] L. Merel, private communication, 1995.

[Mu] R. Murty, A motivated introduction to the Langlands program, in *Advances in Number Theory*, (eds. F. Gouvea and N. Yui), Oxford University Press, 1994.

[MS] V. Kumar Murty and T. Stefanicki, Non-vanishing of quadratic twists of L-functions attached to automorphic representations of $GL(2)$ over $\mathbb{Q}$, preprint, 1994.

[PP] A. Perelli and J. Pomykala, Averages of twisted L-functions, to appear in *Acta Arithmetica*.

[PS] I. Piatetski-Shapiro, The work of Waldspurger, in *Springer Lecture Notes*, **1041** pp. 280–302.

[R1] D. Rohrlich, Non-vanishing of L-functions for $GL(2)$, *Inventiones Math.*, **97** (1989) pp. 381–403.

[R2] D. Rohrlich, Non-vanishing of L-functions and the structure of Mordell-Weil groups, *J. reine angew. Math.*, **417** (1991) pp. 1–26.

[St] T. Stefanicki, Non-vanishing of L-functions attached to automorphic representations of $GL(2)$, Ph.D. Thesis, McGill University, 1992.

[Z] Y. Zhang, Some analytic properties of automorphic L-functions, Ph.D. Thesis, McGill University, 1994.

Name Index

Subject Index